UNITED STATES DEPARTMENT OF DEFENSE
FISCAL YEAR 2012 BUDGET REQUEST

PROGRAM ACQUISITION COSTS BY WEAPON SYSTEM

February 2011

OFFICE OF THE UNDER SECRETARY OF DEFENSE (COMPTROLLER) / CFO

FY 2012 Program Acquisition Costs by Weapon System

Major Weapon Systems

OVERVIEW

The combined capabilities and performance of U.S. weapons systems are unmatched throughout the world, ensuring that our military forces have the advantage over any adversary. The DoD (FY) 2012 request totals $553.1 billion, of which $203.8 billion is for Procurement, and Research, Development, Test and Evaluation (RDT&E) programs. The funding in FY 2012 includes both Base ($188.4 billion) and Overseas Contingency Operations (OCO) ($15.4 billion) funding. For RDT&E appropriations: $75.7 billion (Base, $75.3 billion; OCO, $0.4 billion); for Procurement : $128.1 billion (Base, $113.1 billion; OCO, $15.0 billion). Of this amount, $85.3 billion is for programs that have been designated as Major Defense Acquisition Programs (MDAP). To simplify the display of the various weapon systems, this book is organized by mission area categories.

Funding Categories

- Aircraft
- Command, Control, Communications, and Computer (C4) Systems
- Ground Programs
- Missile Defense
- Munitions and Missiles
- Shipbuilding and Maritime Systems
- Space Based and Related Systems
- Mission Support*
- Science and Technology*

FY 2012 Modernization – Base and OCO: $203.8 Billion

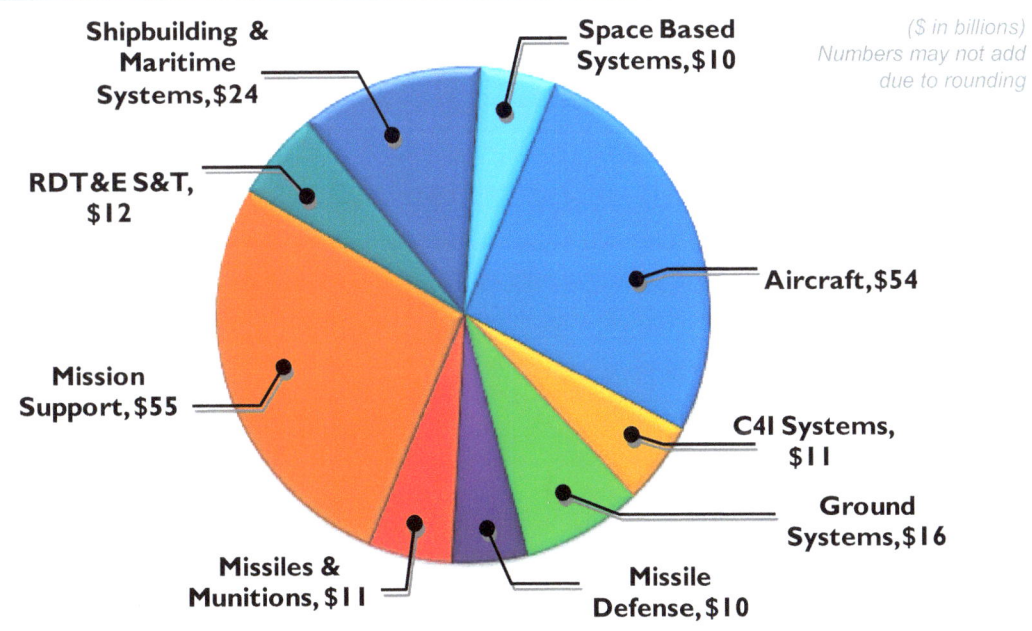

($ in billions)
Numbers may not add due to rounding

Source: FY 2012 PRCP – Investment Categorization

INTRODUCTION

Preparation of this study/report cost the Department of Defense a total of approximately $20,000 for the 2011 Fiscal Year.

Generated on 2011Feb01 1755   RefID: F-C8E6884

FY 2012 Program Acquisition Costs by Weapon System

Major Weapon Systems Summary ($ in Millions)		FY 2010	FY 2011	2012 Base	OCO	Total Request	Page
Aircraft – Joint Service							
MQ–1/MQ–9	Predator and Reaper	1,598.6	2,708.7	2,516.7	10.8	**2,527.5**	1-2
RQ–4	Global Hawk	1,679.1	1,900.3	1,635.4	-	**1,635.4**	1-3
RQ–7/RQ–11	Shadow and Raven	933.2	683.5	246.2	94.6	**340.8**	1-4
C–130J	Hercules	1,192.4	1,450.3	1,257.7	-	**1,257.7**	1-5
JCA	Joint Cargo Aircraft	327.1	377.6	598.8	-	**598.8**	1-6
F–35	Joint Strike Fighter	11,104.5	11,790.0	9,732.8	-	**9,732.8**	1-7
JPATS T–6A	Texan II	278.3	276.7	274.2	-	**274.2**	1-8
V–22	Osprey	2,992.2	2,801.5	2,901.4	70.0	**2,971.4**	1-9
Aircraft – US Army (USA)							
AH–64	Apache Longbow Block 3	411.5	587.1	816.6	-	**816.6**	1-10
CH–47	Chinook	1,058.8	1,180.0	1,409.2	-	**1,409.2**	1-11
LUH	Light Utility Helicopter	325.2	305.3	250.4	-	**250.4**	1-12
UH–60	Black Hawk	1,542.3	1,371.7	1,546.9	72.0	**1,618.9**	1-13
Aircraft – US Air Force (USAF)							
C–17	Globemaster	3,087.7	920.9	527.7	11.0	**538.7**	1-14
KC–X	New Tanker	14.9	863.9	877.1	-	**877.1**	1-15
HH-60M	Pave Hawk	94.9	104.4	104.7	39.3	**144.0**	1-16
F-22	Raptor	838.5	1,238.5	1,063.6	-	**1,063.6**	1-17
Aircraft – US Navy (USN)/US Marine Corps (USMC)							
E–2/D	Advanced Hawkeye	1,126.0	1,132.6	1,222.5	163.5	**1,386.0**	1-18
F/A–18E/F	Super Hornet	1,676.6	1,976.8	2,659.9	2.0	**2,661.9**	1-19
EA–18G	Growler	1,716.8	1,117.1	1,124.6	-	**1,124.6**	1-20
H–1	Huey/Super Cobra	777.3	897.0	841.2	30.0	**871.2**	1-21
MH–60R	Multi-Mission Helicopter	1,002.9	1,161.0	1,018.2	-	**1,018.2**	1-22
MH–60S	Fleet Combat Support Helicopter	519.4	588.8	513.5	-	**513.5**	1-23
P–8A	Poseidon	3,040.8	2,992.3	2,996.5	-	**2,996.5**	1-24
C4 Systems – Joint Service							
JTRS	Joint Tactical Radio System	921.2	967.2	1,541.4	.5	**1,541.9**	2-2
C4 Systems – USA							
E-IBCT	Early-Infantry Brigade Combat Team (E-IBCT) Modernization	2,086.0	2,251.0	749.0	-	**749.0**	2-3
WIN–T	Warfighter Information Network – Tactical	774.6	612.7	1,272.2	.5	**1,272.7**	2-4
Ground Programs – Joint Service							
JTLV	Joint Light Tactical Vehicle	83.9	84.7	243.9	-	**243.9**	3-2
Ground Programs – USA							

Note: FY 2010 and FY 2011 funding includes both Base and OCO. *Numbers may not add due to rounding.*

FY 2012 Program Acquisition Costs by Weapon System

Major Weapon Systems Summary
($ in Millions)

		FY 2010	FY 2011	2012 Base	2012 OCO	2012 Total Request	Page
FHTV	Family Of Heavy Tactical Vehicles	1,410.7	741.9	632.8	47.2	**680.0**	3-3
FMTV	Family Of Medium Tactical Vehicles	1,349.8	1,438.2	436.9	11.1	**448.0**	3-4
M1 Upgrade	Abrams Tank	278.8	290.5	191.0	-	**191.0**	3-5
Stryker	Stryker Family of Armored Vehicles	609.1	435.8	834.0	-	**834.0**	3-6

Missile Defense – Joint Service

		FY 2010	FY 2011	2012 Base	2012 OCO	2012 Total Request	Page
BMD	Ballistic Missile Defense						4-2
BMD	Ballistic Missile Defense	9449.6	10,219.9	10,671.6	-	**10,671.6**	4-3
AEGIS	AEGIS Ballistic Missile Defense	1,644.6	1,561.4	1,525.7	-	**1,525.7**	4-4
THAAD	Terminal High Altitude Area Defense	1,109.1	1,295.4	1,174.7	-	**1,174.7**	4-5
Patriot/PAC–3	Patriot, Army	362.7	498.7	713.2	-	**713.2**	4-6
Patriot/MEADS	Patriot Medium Extended Air Defense System	571.0	467.1	406.6	-	**406.6**	4-7
PAC-3/MSE Missile	PAC-3/MSE Missile	-	62.5	163.9	-	**163.9**	4-8
GMD	Ground-Based Midcourse Defense	1,022.0	1,346.2	1,161.0	-	**1,161.0**	4-9
JLENS	Joint Land Attack Cruise Missile Defense Elevated Netted Sensor	317.1	372.5	344.7	-	**344.7**	4-10
PAA	Phased Adaptive Approach	118.5	441.7	628.4	-	**628.4**	4-11

Munitions and Missiles – Joint Service

		FY 2010	FY 2011	2012 Base	2012 OCO	2012 Total Request	Page
AMRAAM	Advanced Medium Range Air-Air Missile	467.0	577.1	579.5	-	**579.5**	5-2
AIM–9X	Air Intercept Missile - 9X	141.4	127.6	153.5	-	**153.5**	5-3
Chem–Demil	Chemical Demilitarization	1,712.3	1,592.0	1,629.7	-	**1,629.7**	5-4
JAGM	Joint Air-to-Ground Missile	180.3	231.1	245.5	-	**245.5**	5-5
JASSM	Joint Air-to-Surface Standoff Missile	81.0	235.8	242.0	-	**242.0**	5-6
JDAM	Joint Direct Attack Munition	242.3	252.6	76.6	34.1	**110.7**	5-7
JSOW	Joint Standoff Weapon	151.9	143.9	145.4	-	**145.4**	5-8
SDB	Small Diameter Bomb	309.3	332.5	188.0	12.3	**200.3**	5-9

Munitions and Missiles – USA

		FY 2010	FY 2011	2012 Base	2012 OCO	2012 Total Request	Page
Javelin	Javelin Advanced Tank Weapon	258.6	173.9	178.1	-	**178.1**	5-10
GMLRS	Guided Multiple Launch Rocket System (GMLRS)	379.9	342.6	380.8	19.0	**399.8**	5-11

Note: FY 2010 and FY 2011 funding includes both Base and OCO

Numbers may not add due to rounding

FY 2012 Program Acquisition Costs by Weapon System

Major Weapon Systems Summary ($ in Millions)		FY 2010	FY 2011	2012 Base	OCO	Total Request	Page
Munitions and Missiles – USN							
ESSM	Evolved Seasparrow Missile	51.2	48.2	48.5	-	**48.5**	5-12
RAM	Rolling Airframe Missile	69.7	75.0	66.2	-	**66.2**	5-13
Standard	Standard Family of Missiles	338.6	392.1	467.0	-	**467.0**	5-14
Tomahawk	Tactical Tomahawk Cruise Missile	293.2	310.8	312.1	-	**312.1**	5-15
Trident II	Trident II Ballistic Missile	1,114.7	1,188.1	1,398.0	-	**1,398.0**	5-16
Shipbuilding and Maritime Systems – Joint Service							
JHSV	Joint High Speed Vessel	391.1	209.6	416.0	-	**416.0**	6-2
Shipbuilding and Maritime Systems USN							
DDG 51	AEGIS Destroyer	2,483.6	2,970.2	2,081.4	-	**2,081.4**	6-3
LCS	Littoral Combat Ship	1,579.1	1,818.6	2,168.5	-	**2,168.5**	6-4
LPD 17	Amphibious Transport Dock Ship	1,157.8	1.4	1,848.3	-	**1,848.3**	6-5
SSN 774	VIRGINA Class Submarine	4,234.4	5,420.2	4,954.9	-	**4,954.9**	6-6
LHA-7	LHA Replacement	169.5	949.9	2,018.7	-	**2,018.7**	6-7
MLP	Mobile Landing Platform	120.0	380.0	425.9	-	**425.9**	6-8
Space Based and Related Systems – USN							
MUOS	Mobile User Objective System	908.2	911.4	482.4	-	**482.4**	7-2
Space Based and Related Systems – USAF							
AEHF	Advanced Extremely High Frequency	2,292.2	598.4	974.5	-	**974.5**	7-3
EELV	Evolved Expendable Launch Vehicle	1,138.7	1,184.2	1,760.2	-	**1,760.2**	7-4
GPS	Global Positioning System	880.4	1,057.5	1,462.0	-	**1,462.0**	7-5
DWSS	Defense Weather Satellite System	398.9	351.8	444.9	-	**444.9**	7-6
SBIRS	Space Based Infrared System	987.4	1,525.5	995.2	-	**995.2**	7-7
WGS	Wideband Global SATCOM System	279.6	611.8	481.5	-	**481.5**	7-8

Note: FY 2010 and FY 2011 funding includes both Base and OCO

Numbers may not add due to rounding

This page intentionally left blank.

Aircraft

Aviation forces — including fighter/attack, bomber, mobility (cargo/tanker) and specialized support aircraft — provide a versatile striking force capable of rapid deployment worldwide. These forces can quickly gain and sustain air dominance over regional aggressors, permitting rapid air attacks on enemy targets while providing security to exploit the air for logistics, command and control, intelligence, and other functions. Fighter/attack aircraft operate from both land bases and aircraft carriers to combat enemy fighters and attack ground and ship targets. Bombers provide an intercontinental capability to rapidly strike surface targets. The specialized aircraft supporting conventional operations perform functions such as surveillance, airborne warning and control, air battle management, suppression of enemy air defenses, reconnaissance, and combat search and rescue. In addition to these forces, the U.S. military operates a variety of air mobility forces including cargo, aerial-refueling aircraft, helicopters, and support aircraft.

Aircraft funding has continued to increase to support the procurement of aircraft such as the F-35 Joint Strike Fighter, the V-22 Osprey, and the Navy's F/A-18 E/F Super Hornet and E/A-18G Growler.

FY 2012 Aircraft – Base and OCO: $54.2 Billion

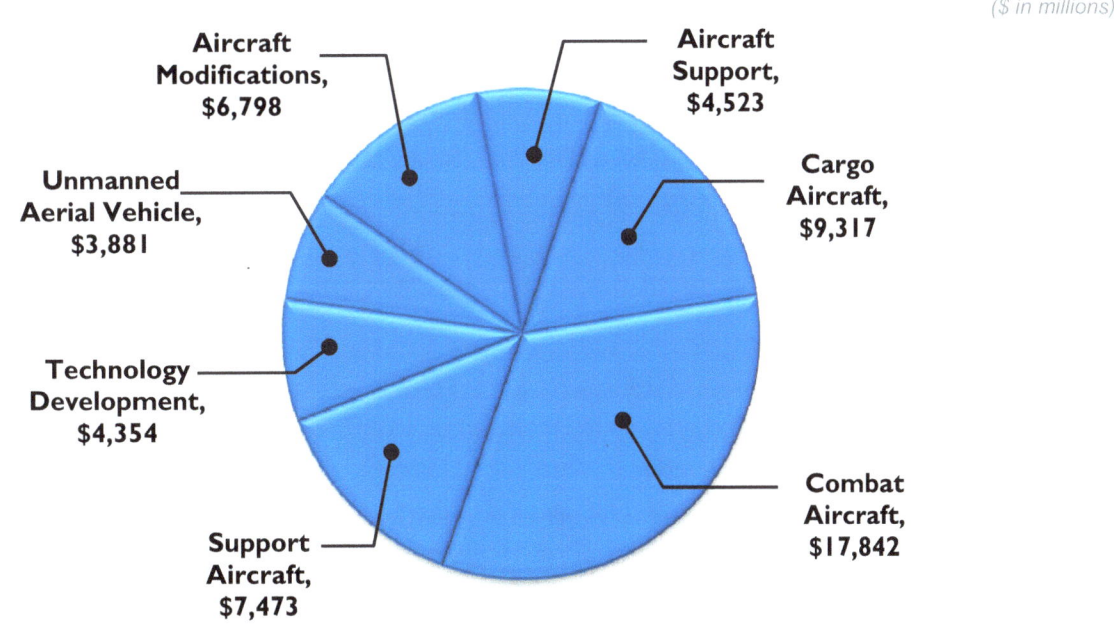

($ in millions)

- Aircraft Modifications, $6,798
- Aircraft Support, $4,523
- Unmanned Aerial Vehicle, $3,881
- Cargo Aircraft, $9,317
- Technology Development, $4,354
- Combat Aircraft, $17,842
- Support Aircraft, $7,473

Source: FY 2012 PRCP – Investment Categorization
Numbers may not add due to rounding

FY 2012 Program Acquisition Costs by Weapon System

MQ-1 Predator/MQ-9 Reaper — DOD - JOINT

USAF Photos

The Predator and Reaper Unmanned Aerial Systems (UASs) are comprised of an aircraft segment consisting of aircraft configured with an array of sensors to include day/night Full Motion Video (FMV), Signals Intelligence (SIGINT), and Synthetic Aperture Radar (SAR) sensor payloads, avionics, and data links; a ground control segment consisting of a Launch and Recovery Element (LRE), and a Mission Control Element (MCE) with embedded Line-of-Sight (LOS) and Beyond-Line-of-Sight (BLOS) communications equipment; a support element; and trained personnel. The Army MQ-1C Gray Eagle has a 2.0L heavy fuel piston engine; where as the Air Force MQ-1B Predator has an aviation fuel piston engine and the Air Force MQ-9 Reaper has a turboprop engine.

Mission: A single-engine, remotely piloted armed reconnaissance aircraft designed to operate over-the-horizon at medium altitude for long endurance. The primary mission is reconnaissance with an embedded strike capability against critical, perishable targets. The Army MQ-1C Gray Eagle also has the unique mission of communications relay.

FY 2012 Program: Continues implementation of transformation towards development and fielding of UASs. Predator and Reaper aircraft support 50 Combat Air Patrols (CAP)/orbits by the end of year FY 2011 and 65 CAPs by the end of FY 2013.

Prime Contractor: General Atomics–Aeronautical Systems Inc., San Diego, CA

MQ-1 Predator/MQ-9 Reaper

	FY 2010*		FY 2011**		FY 2012 Base Budget		FY 2012 OCO Budget		FY 2012 Total Request	
	$M	Qty	$M	Qty	$M	Qty	$M	Qty	$M	Qty
RDT&E										
Predator USAF	23.7	-	28.9	-	14.1	-	-	-	14.1	-
Reaper USAF	104.2	-	125.4	-	146.8	-	-	-	146.8	-
Gray Eagle USA	84.9	-	123.2	-	137.0	-	-	-	137.0	-
Subtotal	212.8	-	277.5	-	297.9	-	-	-	297.9	-
Procurement									-	-
Predator USAF	190.8	-	246	-	158.4	-	-	-	158.4	-
Reaper USAF	573.9	24	1,392.6	48	1,069.3	48	-	-	1,069.3	48
Gray Eagle USA	530.7	24	625.1	26	795.0	36	10.8	-	805.8	36
Subtotal	1,295.4	48	2,263.4	-	2,022.7	84	10.8	-	2,033.5	84
Spares	90.4	-	167.8	-	196.1	-	-	-	196.1	-
Total	1,598.6	48	2,708.7	-	2,516.7	84	10.8	-	2,527.5	84

* FY 2010 & FY 2011 include Base and OCO funding
** Reflects the FY 2011 President's Budget Request
Numbers may not add due to rounding

FY 2012 Program Acquisition Costs by Weapon System

RQ-4 Global Hawk — DOD - JOINT

The RQ-4/MQ-4C unmanned aircraft supports both Air Force and Navy capabilities. The USAF RQ-4 Block 20 includes a communications-relay payload, the Block 30 includes a multi-intelligence suite for imagery and signals intelligence collection, and the Block 40 carries the Multi-Platform Radar Technology Insertion Program for imaging synthetic-aperture radar (SAR) and moving target detection. The USN MQ-4C Broad Area Maritime Surveillance (BAMS) UAS is a tactical asset including payloads for maritime SAR and Inverse SAR, Electro-optical/Infra-red (EO/IR)/Full Motion Video, Electronic Support Measures (ESM), Automatic Identification System (AIS), a basic communications relay capability and Link-16. Each variant features >24 hour endurance and autonomous flight capability. Remote operators control/monitor the aircraft and handle mission planning duties.

Mission: The Air Force RQ-4 performs high-altitude, near-real-time, high-resolution Intelligence, Surveillance, and Reconnaissance (ISR) collection while the Navy MQ-4C provides persistent maritime ISR to Joint, Combatant Commander (COCOM) and Navy numbered Fleet commanders from five orbits worldwide.

FY 2012 Program: Procures three USAF aircraft, payloads, integrated logistics support (to include initial spares, support equipment, technical data, etc.), other support requirements (training devices, etc.), testing, program management support, and related tasks. Also supports continued Navy System Development and Demonstration (SDD).

Prime Contractor: Northrop Grumman Corporation, Rancho Bernardo, CA and Bethpage, NY

RQ-4 Global Hawk

	FY 2010*		FY 2011**		FY 2012 Base Budget		FY 2012 OCO Budget		FY 2012 Total Request	
	$M	Qty	$M	Qty	$M	Qty	$M	Qty	$M	Qty
RDT&E										
RQ-4, USAF	309.2	-	251.3	-	423.5	-	-	-	423.5	-
RQ-4, USN	439.0	2	529.3	-	548.5	-	-	-	548.5	-
Subtotal	748.2	2	780.6	-	972.0	-	-	-	972.0	-
Procurement										
RQ-4, USAF	800.1	4	859.2	4	484.6	3	-	-	484.6	3
Subtotal	800.1	4	859.2	4	484.6	3	-	-	484.6	3
Spares	130.8	-	260.5	-	178.8	-	-	-	178.8	-
Total	1,679.1	6	1,900.3	4	1,635.4	3	-	-	1,635.4	3

* FY 2010 & FY 2011 include Base and OCO funding
** Reflects the FY 2011 President's Budget Request

Numbers may not add due to rounding

FY 2012 Program Acquisition Costs by Weapon System

RQ-7 Shadow/RQ-11 Raven — DOD - JOINT

The RQ-7 and RQ-11 unmanned aircraft are deployable with ground forces that provide tactical Intelligence, Surveillance, and Reconnaissance (ISR).

Mission: The Shadow provides the tactical maneuver commander near-real-time reconnaissance, surveillance, target acquisition, and force protection during day/night and limited adverse weather conditions. Raven is an "over the hill" rucksack-portable, day/night, limited adverse weather, remotely-operated, multi-sensor system in support of combat battalions and below as well as selected combat support units.

FY 2012 Program: Procures multiple variations of quantities for the small unmanned Raven-class aircraft, system hardware, contractor logistics support, and new training equipment for Tactical Common Data Link (TCDL). Additionally, air vehicle modifications and equipment to support Shadow Common Configuration; 16 Laser Designator payload retrofit kits; and 400 one system remote video terminals (OSRVTs).

Prime Contractors: Shadow: AAI Corporation Hunt Valley, MD
Raven: AeroVironment, Monrovia, CA

RQ-7 Shadow/RQ-11 Raven

	FY 2010*		FY 2011**		FY 2012					
					Base Budget		OCO Budget		Total Request	
	$M	Qty	$M	Qty	$M	Qty	$M	Qty	$M	Qty
RDT&E										
Shadow USA	41.9	-	7.8	-	31.9	-	-	-	31.9	-
Shadow USMC	3.1	-	0.9	-	0.9	-	-	-	0.9	-
Raven USA	2.0	-	1.6	-	1.9	-	-	-	1.9	-
Raven USMC	0.6	-	0.5	-	1.0	-	-	-	1.0	-
Subtotal	47.6	-	10.8	-	35.7	-	-	-	35.7	-
Procurement										
Shadow USA	649.9	-	602.8	-	126.2	-	94.6	-	220.8	-
Shadow USMC	109.9	4	26.1	-	11.4	-	-	-	11.4	-
Raven USA	84.3	876	37.6	312	70.8	1,272			70.8	1,272
Raven USMC	41.5	513	6.2	-	2.1	-	-	-	2.1	-
Subtotal	885.6	1,393	672.7	312	210.5	1,272	94.6	-	305.1	1,272
Spares	-	-	-	-	-	-	-	-	-	-
Total	933.2	1,393	683.5	312	246.2	1,272	94.6	-	340.8	1,272

* FY 2010 & FY 2011 include Base and OCO funding
** Reflects the FY 2011 President's Budget Request

Numbers may not add due to rounding

FY 2012 Program Acquisition Costs by Weapon System

C-130J Hercules — DOD - JOINT

USAF Photo

The C-130J Hercules is a tactical airlift aircraft modernizing the U.S. tactical airlift capability. It is capable of performing a number of tactical airlift missions including deployment and redeployment of troops and/or supplies within/between command areas in a theater of operation, aeromedical evacuation, air logistic support and augmentation of strategic airlift forces.

Mission: The mission of the C-130J is the immediate and responsive air movement and delivery of combat troops and supplies directly into objective areas primarily through airlanding, extraction, and airdrop and the air logistic support of all theater forces.

FY 2012 Program: Continues the procurement of C-130J aircraft, by funding one C-130J and ten HC/MC-130s for the Air Force and one KC-130J for the Marine Corps in FY 2012.

Prime Contractor: Lockheed Martin Corporation, Marietta, GA

C-130J Hercules

	FY 2010*		FY 2011**		FY 2012					
					Base Budget		OCO Budget		Total Request	
	$M	Qty	$M	Qty	$M	Qty	$M	Qty	$M	Qty
RDT&E										
HC/MC-130	20.5		15.5		27.1				27.1	-
C-130J	29.1		26.8		39.5				39.5	-
Subtotal	49.6	-	42.3	-	66.6	-	-	-	66.6	-
Procurement	**USAF**									
C-130J	464.4	4	511.3	8	72.9	1			72.9	1
HC/MC-130	511.0	2	896.7	9	1,023.8	10			1,023.8	10
Subtotal	975.4	6	1,408.0	17	1,096.7	11	-	-	1,096.7	11
Procurement	**USN**									
HC-130J***	167.4	2	-	-						
KC-130J	-	-	-	-	87.3	1	-	-	87.3	1
Spares	-	-	-	-	7.1	-	-	-	7.1	-
Total	1,192.4	6	1,450.3	17	1,257.7	12	-	-	1,257.7	12

* FY 2010 & FY 2011 include Base and OCO funding
** Reflects the FY 2011 President's Budget Request
*** FY 2010 Navy funding includes 2 HC-130J aircraft replacements procured on behalf of the US Coast Guard

Numbers may not add due to rounding

FY 2012 Program Acquisition Costs by Weapon System

C-27J Joint Cargo Aircraft

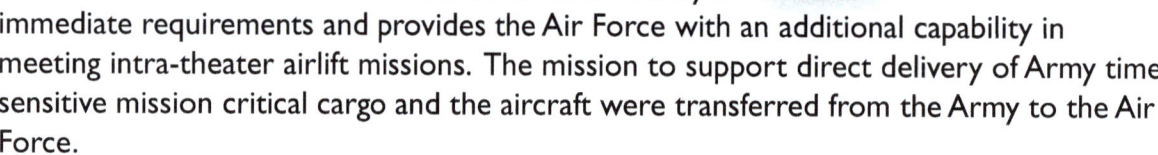

The C-27J Joint Cargo Aircraft (JCA) is an intra-theater light cargo fixed-wing airlift platform that will meet the warfighter need for intra-theater airlift. The aircraft is a commercial derivative aircraft that meets the Army's immediate requirements and provides the Air Force with an additional capability in meeting intra-theater airlift missions. The mission to support direct delivery of Army time sensitive mission critical cargo and the aircraft were transferred from the Army to the Air Force.

Mission: The JCA will provide responsive, flexible, and tailored airlift for combat, humanitarian operations and homeland defense.

FY 2012 Program: Continues procurement of JCA for the Air Force with nine aircraft in FY 2012.

Prime Contractor: L-3 Communications, Greenville, TX

Joint Cargo Aircraft

	FY 2010*		FY 2011**		FY 2012 Base Budget		FY 2012 OCO Budget		FY 2012 Total Request	
	$M	Qty	$M	Qty	$M	Qty	$M	Qty	$M	Qty
RDT&E										
USAF	9.0		26.4		27.1				27.1	-
Subtotal	9.0	-	26.4	-	27.1	-	-	-	27.1	-
Procurement										
USA									-	-
USAF	318.1	8	351.2	8	480.0	9			480.0	9
Subtotal	318.1	8	351.2	8	480.0	9	-	-	480.0	9
Spares	-	-	-	-	91.7	-	-	-	91.7	-
Total	327.1	8	377.6	8	598.8	9	-	-	598.8	9

* FY 2010 & FY 2011 include Base and OCO funding
** Reflects the FY 2011 President's Budget Request

Numbers may not add due to rounding

F-35 Joint Strike Fighter

DOD - JOINT

The F–35 Joint Strike Fighter (JSF) is the next-generation strike fighter for the Navy, Marine Corps, Air Force, and U.S. Allies. The JSF consists of three variants: Conventional Take-Off and Landing (CTOL), Short Take-Off and Vertical Landing (STOVL), and Carrier (CV).

Mission: The JSF will complement the Navy F/A-18E/F and the Air Force F-22 aircraft and will replace the Marine Corps AV-8B, F/A-18C/D and the Air Force A-10 and F-16 aircraft. The JSF will provide all–weather, precision, stealthy, air–to–air and ground strike capability, including direct attack on the most lethal surface–to–air missiles and air defenses.

FY 2012 Program: Restructures the Joint Strike Fighter program to stabilize schedule and cost, scaling back procurement of the Marine Corps' STOVL variant and increasing development to support additional developmental testing. The budget procures 7 CV for Navy, 6 STOVL for Marine Corps and 19 CTOL for Air Force in FY 2012, continuing concurrent aircraft development and production.

Prime Contractors: Lockheed Martin Corporation, Fort Worth, TX
Pratt & Whitney; General Electric/Rolls Royce Fighter Engine Team

F-35 Joint Strike Fighter

	FY 2010*		FY 2011**		FY 2012 Base Budget		FY 2012 OCO Budget		FY 2012 Total Request	
	$M	Qty	$M	Qty	$M	Qty	$M	Qty	$M	Qty
RDT&E										
USN	1,886.2		1,375.7		1,348.2				1,348.2	-
USAF	2,033.5		1,101.3		1,435.8				1,435.8	-
Subtotal	3,919.7	-	2,477.0	-	2,784.0	-	-	-	2,784.0	-
Procurement										
USN	4,449.3	20	4,463.1	20	2,979.9	13			2,979.9	13
USAF	2,357.9	10	4,315.1	23	3,664.1	19			3,664.1	19
Subtotal	6,807.3	30	8,778.2	43	6,644.0	32	-	-	6,644.0	32
Spares	377.5	0	534.7	0	304.8	0	0	0	304.8	0
Total	11,104.5	30	11,790.0	43	9,732.8	32	-	-	9,732.8	32

* FY 2010 & FY 2011 include Base and OCO funding
** Reflects the FY 2011 President's Budget Request and includes $204.9 million OCO funding for 1 combat replacement aircraft for Air Force

Numbers may not add due to rounding

FY 2012 Program Acquisition Costs by Weapon System

JPATS T-6B Texan II — DOD - JOINT

The Joint Primary Aircraft Training System (JPATS) is a joint Navy/Air Force program that will use the T-6B Texan as a replacement for the Services' fleets of primary trainer aircraft (T-34 and T-37, respectively) and associated Ground Based Training Systems. The T-6 Texan II is a tandem seat, turboprop aircraft derivative of the Pilatus PC-9 powered by a single Pratt & Whitney PT6A-68 engine.

Mission: Supports joint Navy and Air Force specialized undergraduate pilot training.

FY 2012 Program: Continues production of JPATS aircraft, supporting procurement of 36 aircraft and associated support for the Navy in FY 2012.

Prime Contractor: Hawker Beechcraft, Wichita, KS

JPATS T-6B Texan II

	FY 2010*		FY 2011**		FY 2012 Base Budget		FY 2012 OCO Budget		FY 2012 Total Request	
	$M	Qty	$M	Qty	$M	Qty	$M	Qty	$M	Qty
RDT&E									-	-
Procurement										
USN	255.4	37	266.1	38	266.9	36			266.9	36
USAF	12.7	-	-	-					-	-
Subtotal	268.1	37	266.1	38	266.9	36	-	-	266.9	36
Spares	10.2	-	10.6	-	7.3	-	-	-	7.3	-
Total	278.3	37	276.7	38	274.2	36	-	-	274.2	36

* FY 2010 & FY 2011 include Base and OCO funding
** Reflects the FY 2011 President's Budget Request

Numbers may not add due to rounding

FY 2012 Program Acquisition Costs by Weapon System

V-22 Osprey

The V-22 Osprey is a tilt-rotor, vertical takeoff and landing aircraft designed to meet the amphibious/vertical assault needs of the Marine Corps, the strike rescue needs of the Navy and long range special operations forces (SOF) missions for US Special Operations Command (USSOCOM). The aircraft is designed to fly 2,100 miles with one in-flight refueling, giving the services the advantage of a vertical and/or short takeoff and landing (V/STOL) aircraft that could rapidly self-deploy to any location in the world.

Mission: The V-22 mission includes airborne assault, vertical lift, combat search and rescue, and special operations.

FY 2012 Program: Supports procurement of 30 MV-22 aircraft for the Navy and 5 CV-22 aircraft for USSOCOM, and one additional Air Force aircraft in OCO to replace a combat loss. The procurement objective is 458 aircraft (408 MV-22 aircraft for the Navy/Marine Corps and 50 CV-22 aircraft for USSOCOM). The program is being executed under a 5-year multiyear procurement contract, which began in FY 2008.

Prime Contractor: Bell Helicopter, Fort Worth, TX

V-22 Osprey

	FY 2010*		FY 2011**		FY 2012 Base Budget		FY 2012 OCO Budget		FY 2012 Total Request	
	$M	Qty	$M	Qty	$M	Qty	$M	Qty	$M	Qty
RDT&E										
USN	79.9	-	46.1	-	84.5	-			84.5	-
USAF	19.0	-	18.3	-	20.7	-			20.7	-
Subtotal	98.9	-	64.4	-	105.2	-	-	-	105.2	-
Procurement										
USN	2,284.9	30	2,202.9	30	2,308.8	30			2,308.8	30
USAF	449.7	5	406.7	5	359.9	5	70.0	1	429.9	6
Subtotal	2,734.6	35	2,609.6	35	2,668.7	35	70.0	1	2,738.7	36
USN	-	-	18.9	-	-	-	-	-	-	-
USAF	116.5	-	108.6	-	57.5	-	-	-	57.5	-
Spares	158.7	-	127.5	-	127.5	-	-	-	127.5	-
USN Subtotal	2,364.8	30	2,267.9	30	2,393.3	30	-	-	2,393.3	30
USAF Subtotal	585.2	5	533.6	5	438.1	5	-	-	438.1	5
Total	2,992.2	35	2,801.5	35	2,901.4	35	70.0	1	2,971.4	36

* FY 2010 & FY 2011 include Base and OCO funding
** Reflects the FY 2011 President's Budget Request

Numbers may not add due to rounding

FY 2012 Program Acquisition Costs by Weapon System

AH-64 Apache Block 3: New Build/ReManufacture

US Army Photo

The Apache Block 3 program consists of a mast mounted Fire Control Radar (FCR) integrated into an upgraded and enhanced AH–64 airframe. This program also provides for the installation of the Target Acquisition Designation Sight (TADS) and Pilot Night Vision Sensors (PNVS), plus other safety and reliability enhancements.

Mission: The AH–64 provides a fire-and-forget HELLFIRE air-to-ground missile capability, modernized target acquisition and night vision capabilities.

FY 2012 Program: Supports the remanufacture of 19 AH-64 aircraft to the AH-64 D (Longbow) Block 3 configuration. The AH-64 Block 3 program is comprised of both remanufactured and new build aircraft. The first new build aircraft will be funded in the FY 2013 program, with long lead funds included in the FY 2012 request.

Prime Contractors: Integration: Northrop Grumman Corporation, Baltimore, MD
　　　　　　　　　　　　　　　Lockheed Martin Corporation, Oswego, NY
　　　　　　　　Apache: The Boeing Company, Mesa, AZ

AH–64 Apache Block 3: New Build/ReManufacture

	FY 2010*		FY 2011**		FY 2012***					
					Base Budget		OCO Budget		Total Request	
ReManufacture AB3A	$M	Qty	$M	Qty	$M	Qty	$M	Qty	$M	Qty
RDT&E	146.9	-	93.3	-	92.8	-			92.8	-
Procurement	264.6	8	493.8	16	619.6	19			619.6	19
Spares	-	-	-	-	-	-	-	-	-	-
Total	411.5	8	587.1	16	712.4	19	-	-	712.4	19
New Build AB3B	$M	Qty	$M	Qty	$M	Qty	$M	Qty	$M	Qty
RDT&E	-	-	-	-	-	-			-	-
Procurement	-	-	-	-	104.2	-			104.2	-
Spares	-	-	-	-	-	-			-	-
Total	-	-	-	-	104.2	-	-	-	104.2	-
Grand Totals	411.5	8	587.1	16	816.6	19	-	-	816.6	19

* FY 2010 & FY 2011 include Base and OCO funding
** Reflects the FY 2011 President's Budget Request
*** FY 2012 OCO request excludes 1 Block-2 War Replacement Aircraft ($35.5M)

Numbers may not add due to rounding

FY 2012 Program Acquisition Costs by Weapon System

CH–47 Chinook

US Army Photo

The CH-47F program procures new and remanufactured/Service Life Extension Program CH-47F helicopters. The aircraft include an upgraded digital cockpit and modifications to the airframe to reduce vibration. The upgraded cockpit includes a digital data bus that permits installation of enhanced communications and navigation equipment for improved situational awareness, mission performance, and survivability. The new aircraft uses more powerful T55-GA-714A engines that improve fuel efficiency and enhance lift performance.

Mission: To provide a system designed to transport ground forces, supplies, ammunition, and other battle-critical cargo in support of worldwide combat and contingency operations.

FY 2012 Program: Funds the acquisition of 47 aircraft, of which 32 will be new build aircraft and 15 will be remanufactured/Service Life Extension Program aircraft.

Prime Contractor: The Boeing Company, Philadelphia PA

CH–47F Chinook

	FY 2010*		FY 2011**		FY 2012 Base Budget		FY 2012 OCO Budget		FY 2012 Total Request	
	$M	Qty	$M	Qty	$M	Qty	$M	Qty	$M	Qty
RDT&E	21.5	-	21.0	-	48.9	-			48.9	-
Procurement	1,037.3	37	1,159.0	40	1,360.3	47			1,360.3	47
Spares	-	-	-	-	-	-			-	-
Total	1,058.8	37	1,180.0	40	1,409.2	47	-	-	1,409.2	47

* FY 2010 & FY 2011 include Base and OCO funding
** Reflects the FY 2011 President's Budget Request

Numbers may not add due to rounding

FY 2012 Program Acquisition Costs by Weapon System

LUH Light Utility Helicopter

US Army Image

The Light Utility Helicopter (LUH) will be a utility helicopter replacing the UH-1 and the OH-58 Kiowa Warrior. It will provide reliable and sustainable general and administrative support in permissive environments at reduced acquisition and operating costs. There is no RDT&E funding required for this program. The LUH acquisition strategy is based on a competitive procurement of a commercial-off-the-shelf, non-developmental aircraft.

The UH-72A Lakota is a United States Army light utility helicopter that entered service in 2006. The Lakota is a militarized version of the Eurocopter EC145 modified to an LUH configuration. In June 2006, the US Army selected it as the winner of its LUH program with a 345 aircraft fleet planned.

Mission: The Light Utility Helicopter will provide organic general support at Corps and Division levels. The primary mission for the LUH is to provide aerial transport for logistical and administrative support.

FY 2012 Program: Supports the continued production of 39 aircraft.

Prime Contractor: EADS North America American Eurocopter, Columbus, MS

LUH Light Utility Helicopter

| | FY 2010 | | FY 2011 | | FY 2012 | | | | | |
| | | | | | Base Budget | | OCO Budget | | Total Request | |
	$M	Qty	$M	Qty	$M	Qty	$M	Qty	$M	Qty
RDT&E	-	-	-	-	-	-			-	-
Procurement	325.2	54	305.3	50	250.4	39			250.4	39
Spares	-	-	-	-	-	-			-	-
Total	325.2	54	305.3	50	250.4	39	-	-	250.4	39

* FY 2010 & FY 2011 include Base and OCO funding
** Reflects the FY 2011 President's Budget Request

Numbers may not add due to rounding

FY 2012 Program Acquisition Costs by Weapon System

UH-60 Black Hawk — USA

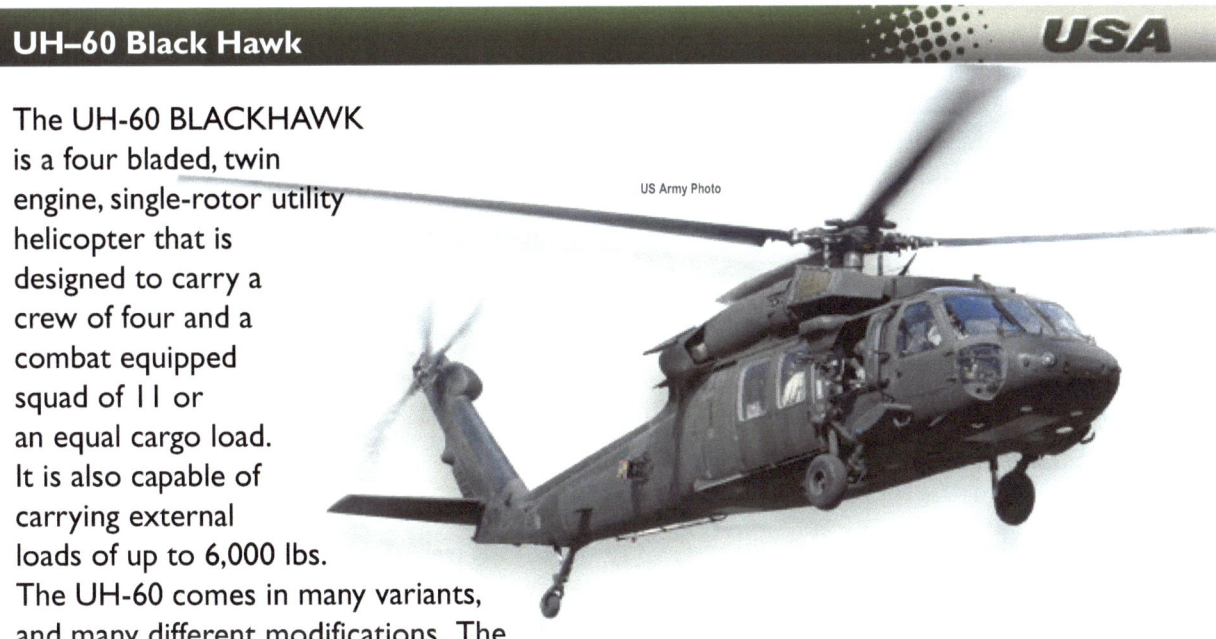

US Army Photo

The UH-60 BLACKHAWK is a four bladed, twin engine, single-rotor utility helicopter that is designed to carry a crew of four and a combat equipped squad of 11 or an equal cargo load. It is also capable of carrying external loads of up to 6,000 lbs. The UH-60 comes in many variants, and many different modifications. The Army variants can be fitted with the stub wings to carry additional fuel tanks or weapons. Variants may have different capabilities and equipment in order to fulfill different roles. The Black Hawk series of aircraft can perform a wide array of missions, including the tactical transport of troops, electronic warfare, and aeromedical evacuation.

Mission: The BLACKHAWK provides a highly maneuverable, air transportable, troop carrying helicopter for all intensities of conflict, without regard to geographical location or environmental conditions. It moves troops, equipment and supplies into combat and performs aeromedical evacuation and multiple functions in support of the Army's air mobility doctrine for employment of ground forces.

FY 2012 Program: The request supports a follow-on 5-year multiyear procurement (MYP) contract for FYs 2012-2016. The program is currently on schedule and within budget. The FY 2012 budget request supports continued production of 75 aircraft; 71 in the base request and 4 in the OCO request to replace combat losses. Specific UH-60 variants funded include the Utility UH model and the Medical HH model.

Prime Contractor: Sikorsky Aircraft, Stratford, CT

UH-60 Black Hawk

	FY 2010*		FY 2011**		FY 2012					
					Base Budget		OCO Budget		Total Request	
	$M	Qty	$M	Qty	$M	Qty	$M	Qty	$M	Qty
RDT&E	59.1	-	20.6	-	21.5	-			21.5	-
Procurement	1,483.2	81	1,351.1	72	1,525.4	71	72.0	4	1,597.4	75
Spares	-	-	-	-	-	-	-	-	-	-
Total	1,542.3	81	1,371.7	72	1,546.9	71	72.0	4	1,618.9	75

* FY 2010 & FY 2011 include Base and OCO funding
** Reflects the FY 2011 President's Budget Request

Numbers may not add due to rounding

FY 2012 Program Acquisition Costs by Weapon System

C-17 Globemaster

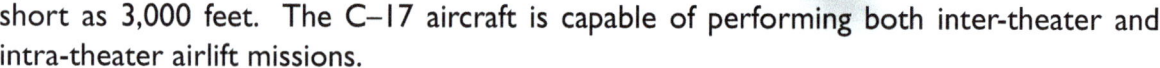

The C-17 Globemaster is a wide-body aircraft capable of airlifting outsized and oversized payloads over intercontinental ranges, with or without in-flight refueling. Its capabilities include rapid direct delivery of forces by airland or airdrop into austere tactical environments with runways as short as 3,000 feet. The C-17 aircraft is capable of performing both inter-theater and intra-theater airlift missions.

USAF Photo

Mission: The C-17 aircraft provides outsize intra-theater airland/airdrop capability not available in the current airlift force. It provides rapid strategic delivery of troops and all types of cargo to main operating bases or directly to forward bases in the deployment area.

FY 2012 Program: Funds modifications to existing C-17 aircraft and continued development and testing of C-17 aircraft performance improvements/mandates and aeromedical evacuation equipment in support of OCO. Supports transition to sustainment in preparation for shutdown activities for production of new aircraft. The Department has determined that the C-17 aircraft already procured are sufficient to satisfy the Department's airlift requirement.

Prime Contractors: The Boeing Company, Long Beach, CA
Pratt & Whitney Corporation, East Hartford, CT

C-17 Globemaster

	FY 2010*		FY 2011**		FY 2012					
					Base Budget		OCO Budget		Total Request	
	$M	Qty	$M	Qty	$M	Qty	$M	Qty	$M	Qty
RDT&E	156.2		177.2		128.2				128.2	-
Procurement	2,931.5	10	743.7		385.9		11.0		396.8	-
Spares					13.7				13.7	-
Total	3,087.7	10	920.9	-	527.7	-	11.0	-	538.7	-

* FY 2010 & FY 2011 include Base and OCO funding
** Reflects the FY 2011 President's Budget Request

Numbers may not add due to rounding

AIRCRAFT

FY 2012 Program Acquisition Costs by Weapon System

KC-X New Tanker

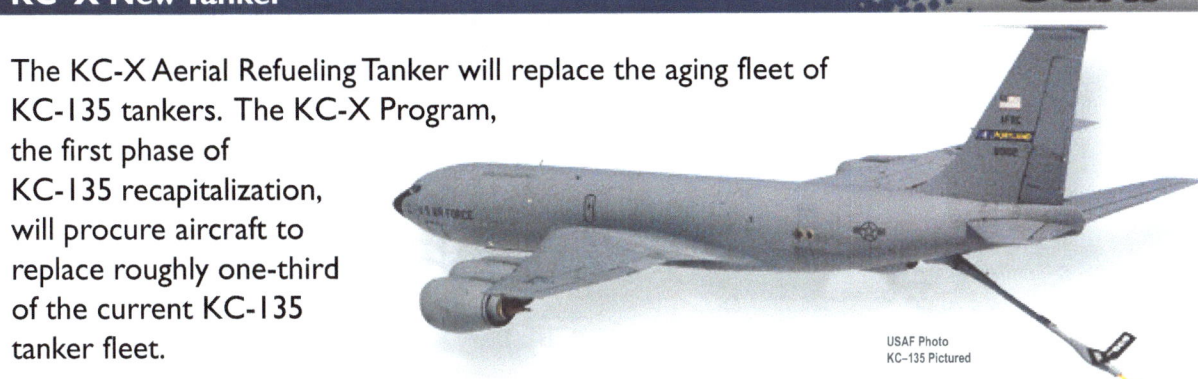

USAF Photo
KC-135 Pictured

The KC-X Aerial Refueling Tanker will replace the aging fleet of KC-135 tankers. The KC-X Program, the first phase of KC-135 recapitalization, will procure aircraft to replace roughly one-third of the current KC-135 tanker fleet.

Mission: The KC-X will meet the primary air refueling missions of Global Attack, Air Bridge, Theater Support, Deployment, and Special Operations Support. Air refueling forces perform these missions at the strategic, operational, and tactical level across the entire spectrum of military operations. Other missions include emergency air refueling, aero medical evacuation, and combat search and rescue.

FY 2012 Program: Continues support for development of the next generation aerial refueling tanker. Source selection for the development contract is pending.

Prime Contractor: To Be Determined.

KC-X New Tanker

| | FY 2010* | | FY 2011** | | FY 2012 | | | | | |
| | | | | | Base Budget | | OCO Budget | | Total Request | |
	$M	Qty	$M	Qty	$M	Qty	$M	Qty	$M	Qty
RDT&E	14.9		863.9		877.1				877.1	-
Procurement									-	-
Spares									-	-
Total	14.9	-	863.9	-	877.1	-	-	-	877.1	-

* FY 2010 & FY 2011 include Base and OCO funding
** Reflects the FY 2011 President's Budget Request

Numbers may not add due to rounding

AIRCRAFT

FY 2012 Program Acquisition Costs by Weapon System

HH–60M Pave Hawk

The HH-60M Pave Hawk is a search and Rescue version of the Army's UH-60M Blackhawk helicopter. The HH-60M is a four bladed, twin engine, single-rotor helicopter that is designed to carry a crew of four and a combat equipped squad of 11 or an equal cargo load. It is also capable of carrying external loads of up to 6,000 lbs. The HH-60M comes in many variants and many different modifications. The Air Force variant can be fitted with the stub wings to carry additional fuel tanks or weapons. Variants may have different capabilities and equipment in order to fulfill different roles. The Pave Hawk series of aircraft can perform a wide array of missions, including the tactical transport of troops, electronic warfare, and aero medical evacuation. The HH-60M will replace the HH-60G.

Mission: The *HH-60* Pave Hawk is the US Air Force version of the UH-60 Black Hawk of the US Army, modified for aircrew *search* and *rescue* in all weather situations. The Pave Hawk perform special missions including search and rescue, combat support, and aero medical evacuation.

FY 2012 Program: Procures 4 aircraft; 3 in the base, and 1 additional aircraft in OCO to replace a combat loss. The budget request supports near term recapitalizing the HH-60G Combat Support and Rescue (CSAR) fleet due to delays in earlier CSAR replacement efforts. Platform configuration is dependent upon finalizing Air Force acquisition plans to acquire a long term replacement CSAR platform.

Prime Contractor: Sikorsky Aircraft, Stratford, CT

HH-60M Pave Hawk

	FY 2010*		FY 2011**		FY 2012					
					Base Budget		OCO Budget		Total Request	
	$M	Qty	$M	Qty	$M	Qty	$M	Qty	$M	Qty
RDT&E	-	-	-	-	-	-			-	-
Procurement	94.9	4	104.4	3	104.7	3	39.3	1	144.0	4
Spares	-	-	-	-	-	-			-	-
Total	94.9	4	104.4	3	104.7	3	39.3	1	144.0	4

* FY 2010 & FY 2011 include Base and OCO funding
** Reflects the FY 2011 President's Budget Request

Numbers may not add due to rounding

FY 2012 Program Acquisition Costs by Weapon System

F-22 Raptor

The F-22 Raptor program is producing the next generation air superiority fighter for the first part of the century. The F-22A will penetrate enemy airspace and achieve first-look, first-kill capability against multiple targets. It has unprecedented survivability and lethality, ensuring the Joint Forces have freedom from attack, freedom to maneuver, and freedom to attack.

Mission: The F-22 will provide enhanced U.S. air superiority capability against the projected threat and will eventually replace the F-15 aircraft.

FY 2012 Program: Supports procurement of equipment associated with standing up operational locations and other support required to deliver new aircraft and funds shutdown activities, preserving assets for long-term F-22 fleet sustainment. Continues critical F-22 modernization through incremental capability upgrades and key reliability and maintainability efforts. Continues retrofit of Increment 3.1 into the combat-coded F-22 fleet. Increment 3.1 provides an initial ground attack kill chain capability via inclusion of emitter-based geo-location of threat systems, ground-looking synthetic aperture radar (SAR) modes, electronic attack capability, and initial integration of the Small Diameter Bomb (SDB-1), which expands the F-22's ground attack arsenal from one Joint Direct Attack Munition (JDAM) to four SDB-1s per payload. Continues development of Increment 3.2, providing AIM-120D and AIM-9X integration, radar electronic protection, enhanced speed and accuracy of target geo-location, Link-16 track fusion, Automatic Ground-Collision Avoidance System (AGCAS), and other enhancements to improve system safety and effectiveness.

Prime Contractors: Lockheed Martin, Marietta, GA and Fort Worth, TX;
Boeing, Seattle, WA;
Pratt & Whitney, West Palm Beach, FL

F-22 Raptor

	FY 2010*		FY 2011**		FY 2012					
					Base Budget		OCO Budget		Total Request	
	$M	Qty	$M	Qty	$M	Qty	$M	Qty	$M	Qty
RDT&E	559.5		576.3		718.4				718.4	-
Procurement	271.7		650.2		336.2				336.2	-
Spares	7.3		11.9		9.0				9.0	-
Total	838.5	-	1,238.5	-	1,063.6	-	-	-	1,063.6	-

* FY 2010 & FY 2011 include Base and OCO funding
** Reflects the FY 2011 President's Budget Request

Numbers may not add due to rounding

FY 2012 Program Acquisition Costs by Weapon System

E-2D Advanced Hawkeye

The E-2D Advanced Hawkeye (AHE) is an airborne early warning, all weather, twin-engine, carrier-based aircraft designed to extend task force defense perimeters. The Advanced Hawkeye provides improved battle space target detection and situational awareness, especially in the littorals; supports the Theater Air and Missile Defense operations; and improves Operational Availability for the radar system.

Mission: The E-2D AHE provides advance warning of approaching enemy surface units and aircraft to vector interceptors or strike aircraft to attack. It provides area surveillance, intercept, strike/air traffic control, radar surveillance, search and rescue assistance, communication relay and automatic tactical data exchange. The E-2D Advanced Hawkeye provides a two-generational leap in radar technology, and will provide the long range air and surface picture; theater air and missile defense, and is a key component of Naval Integrated Fire Control-Counter Air (NIFC-CA).

FY 2012 Program: Funds 5 E-2D AHE Low Rate Initial Production (LRIP) aircraft, associated support, and funds advance procurement for 7 FY 2013 aircraft. Additionally, FY 2012 Overseas Contingency Operations funding for 1 E-2D is requested to replace an FY 2010 combat loss. Research and development funding supports developmental flight testing, pilot production verification and validation activities, trainers, and Mode 5/S non-recurring engineering.

Prime Contractors: Airframe: Northrop Grumman Corporation, Bethpage, NY (Engineering) and St. Augustine, FL (Manufacturing)
Engine: Rolls-Royce Corporation, Indianapolis, IN
Radar: Lockheed Martin Corporation, Syracuse, NY

E-2D Advanced Hawkeye

	FY 2010*		FY 2011**		FY 2012 Base Budget		FY 2012 OCO Budget		FY 2012 Total Request	
	$M	Qty	$M	Qty	$M	Qty	$M	Qty	$M	Qty
RDT&E	346.2	-	171.1	-	111.0	-	-	-	111.0	-
Procurement	742.1	3	937.8	4	1,072.8	5	163.5	1	1,236.3	6
Spares	37.8	-	23.6	-	38.7	-	-	-	38.7	-
Total	1,126.0	3	1,132.6	4	1,222.5	5	163.5	1	1,386.0	6

* FY 2010 & FY 2011 include Base and OCO funding
** Reflects the FY 2011 President's Budget Request

Numbers may not add due to rounding

FY 2012 Program Acquisition Costs by Weapon System

F/A-18E/F Super Hornet — USN

US Navy Photo

The F/A-18E/F Super Hornet is a carrier-based, twin-engine, high-performance, multi-mission, tactical fighter and attack aircraft. With its selected external equipment the aircraft can be optimized to accomplish both fighter and attack missions. The F/A-18E/F provides a 40 percent increase in combat radius, 50 percent increase in endurance, 25 percent greater weapons payload, three times more ordnance, and is five times more survivable than the F/A-18A/C models. The planned spiral developments will allow the program to continue to pace the assessed threat beyond 2025.

Mission: The F/A-18E/F strike fighter performs traditional missions of fighter escort and fleet air defense, interdiction, and close air support, while still retaining excellent fighter and self-defense capabilities. The F/A-18E/F aircraft was designed to replace the F-14 fighter aircraft.

FY 2012 Program: Funds the continued multiyear procurement of 28 F/A-18E/F aircraft, associated spares, and provides the advance procurement resources for 28 FY 2013 aircraft. Continues the research, development, and testing of the planned spiral developments of the F/A-18E/F related systems. Common shared cost between the EA-18G and F/A-E/F programs are funded out of the F/A-E/F program.

Prime Contractors: Airframe: The Boeing Company, St. Louis, MO
Engine: General Electric Aviation, Lynn, MA

F/A-18E/F Super Hornet

	FY 2010*		FY 2011**		FY 2012 Base Budget		FY 2012 OCO Budget		FY 2012 Total Request	
	$M	Qty	$M	Qty	$M	Qty	$M	Qty	$M	Qty
RDT&E	114.1	-	148.4	-	151.0	-	2.0	-	153.0	-
Procurement	1,551.1	18	1,787.2	22	2,431.7	28	-	-	2,431.7	28
Spares	11.3	-	41.2	-	77.2	-	-	-	77.2	-
Total	1,676.6	18	1,976.8	22	2,659.9	28	2.0	-	2,661.9	28

* FY 2010 & FY 2011 include Base and OCO funding
** Reflects the FY 2011 President's Budget Request
Does not include funding for the Service Life Assessment Program (SLAP)

Numbers may not add due to rounding
No modification funding included

FY 2012 Program Acquisition Costs by Weapon System

EA-18G Growler

US Navy Photo

The EA-18G Growler is a tandem two-seat, twin turbojet engine, carrier-based, electronic attack variant of the F/A-18F Super Hornet strike fighter.

Mission: The EA-18G Growler is the first electronic warfare aircraft produced in more than 35 years. The EA-18G provides one of the most flexible offensive Electronic Warfare (EW) capabilities available to the Joint warfighter across the spectrum of conflict from Irregular Warfare to Major Contingency Operations. The EA-18G supports naval, joint, and coalition strike aircraft, providing radar and communications jamming and kinetic effects to increase the survivability and lethality of all strike aircraft. The EA-18G can operate autonomously or as a major node in a network centric operation. The EA-18G's electronic suite can both detect, identify, and locate emitters; and suppress hostile emitters through jamming and kinetic effects. The EA-18G aircraft is built to replace the EA-6B Prowler aircraft.

FY 2012 Program: Funds 12 EA-18G aircraft, associated spares, and provides the advance procurement resources for 12 FY 2013 aircraft. Continues the research, development, and testing of electronic systems and techniques. The aircraft are required to recapitalize the four Navy expeditionary EA-6B squadrons that had been planned to disestablish by the end of FY 2012.

Prime Contractors: Airframe: The Boeing Company, St. Louis, MO
Engine: General Electric Aviation, Lynn, MA

EA-18G Growler

	FY 2010*		FY 2011**		FY 2012					
					Base Budget		OCO Budget		Total Request	
	$M	Qty	$M	Qty	$M	Qty	$M	Qty	$M	Qty
RDT&E	55.5	-	22.0	-	17.1	-	-	-	17.1	-
Procurement	1,627.3	22	1,083.9	12	1,107.5	12	-	-	1,107.5	12
Spares	34.0	-	11.2	-	-	-	-	-	-	-
Total	1,716.8	22	1,117.1	12	1,124.6	12	-	-	1,124.6	12

* FY 2010 & FY 2011 include Base and OCO funding
** Reflects the FY 2011 President's Budget Request

Numbers may not add due to rounding
No modification funding included

FY 2012 Program Acquisition Costs by Weapon System

H-1 Huey/Super Cobra

The H-1 Helicopter Upgrade program converts AH-1W and UH-1N helicopters to the AH-1Z and UH-1Y, respectively. The upgraded helicopters will have increased maneuverability, speed, and payload capability. The upgrade scope includes a new four-bladed rotor system, new transmissions, a new four-bladed tail rotor and drive system, and upgraded landing gear.

Mission: The H-1 Upgrades provide offensive air support, utility support, armed escort, and airborne command and control during naval expeditionary operations or joint and combined operations.

FY 2012 Program: Provides for the production of 25 aircraft (18 UH-1Y new build aircraft, 2 AH-1Z remanufactured aircraft, and 5 new build AH-1Z aircraft). In addition, the request provides for one additional new build AH-1Z aircraft in OCO to replace a combat loss.

Prime Contractor: Bell Helicopter, Fort Worth, TX

H-1 Huey/Super Cobra Upgrades

	FY 2010*		FY 2011**		FY 2012 Base Budget		FY 2012 OCO Budget		FY 2012 Total Request	
	$M	Qty	$M	Qty	$M	Qty	$M	Qty	$M	Qty
RDT&E	31.3	-	60.5	-	72.6				72.6	-
Procurement	746.0	25	808.1	28	768.6	25	30.0	1	798.6	26
Spares	-	-	28.4	-					-	-
Total	777.3	25	897.0	28	841.2	25	30.0	1	871.2	26

* FY 2010 & FY 2011 include Base and OCO funding
** Reflects the FY 2011 President's Budget Request

Numbers may not add due to rounding

FY 2012 Program Acquisition Costs by Weapon System

MH–60R Multi-Mission Helicopter

US Navy Photo

The MH–60R Multi-Mission Helicopter Upgrade program provides battle group protection, and adds significant capability in coastal littorals and regional conflicts. The upgrade includes new H–60 series airframes, significant avionics improvements, enhancements to the acoustic suite, new radars, and an improved electronics surveillance system.

Mission: The MH-60R will be the forward deployed fleet's primary Anti-Submarine and Anti- Surface Warfare platform.

FY 2012 Program: The FY 2012 budget request is for 24 helicopters as part of a follow-on 5 year multiyear procurement (MYP) for MH-60 airframes, from FYs 2012 to 2016. In addition, the request includes funds for a MYP of MH-60 cockpits and sensors for the same period. The Army serves as the executive agent to execute the UH-60 and MH-60 airframe MYP efforts.

Prime Contractors: Airframe: Sikorsky Aircraft, Stratford, CT
Avionics: Lockheed Martin Corporation, Owego, NY

MH–60R Multi-Mission Helicopter

	FY 2010*		FY 2011**		FY 2012					
					Base Budget		OCO Budget		Total Request	
	$M	Qty	$M	Qty	$M	Qty	$M	Qty	$M	Qty
RDT&E	69.4	-	55.8	-	17.7	-			17.7	-
Procurement	931.7	24	1,059.9	24	1,000.5	24			1,000.5	24
Spares	1.8	-	45.3	-	-	-			-	-
Total	1,002.9	24	1,161.0	24	1,018.2	24	-	-	1,018.2	24

* FY 2010 & FY 2011 include Base and OCO funding
** Reflects the FY 2011 President's Budget Request

Numbers may not add due to rounding

MH–60S Fleet Combat Support Helicopter

The MH-60S is a versatile twin-engine helicopter used to maintain forward deployed fleet sustainability through rapid airborne delivery of materials and personnel, to support amphibious operations through search and rescue coverage and to provide an organic airborne mine countermeasures capability.

Mission: The MH-60S will conduct vertical replenishment (VERTREP), day/night ship-to-ship, ship-to-shore, and shore-to-ship external transfer of cargo; internal transport of passengers, mail and cargo, vertical onboard delivery; air operations; and day/night search and rescue. Organic Airborne Mine Countermeasures (OAMCM) has been added as a primary mission for the MH-60S. Five separate sensors will be integrated into the MH 60S helicopter and will provide Carrier Battle Groups and Amphibious Readiness Groups with an OAMCM capability.

FY 2012 Program: The FY 2012 budget request is for 18 helicopters as part of a follow-on 5 year multiyear procurement (MYP) for MH-60 airframes, from FYs 2012 to 2016. In addition, the request includes funds for a MYP of MH-60 cockpits and sensors for the same period. The Army serves as the executive agent to execute the UH-60 and MH-60 airframe MYP efforts.

Prime Contractor: Sikorsky Aircraft, Stratford, CT

MH–60S Fleet Combat Support Helicopter

	FY 2010*		FY 2011**		FY 2012 Base Budget		FY 2012 OCO Budget		FY 2012 Total Request	
	$M	Qty	$M	Qty	$M	Qty	$M	Qty	$M	Qty
RDT&E	47.9	-	38.9	-	30.6	-			30.6	-
Procurement	471.5	18	548.7	18	482.9	18			482.9	18
Spares	-	-	1.2	-	-	-			-	-
Total	519.4	18	588.8	18	513.5	18	-	-	513.5	18

* FY 2010 & FY 2011 include Base and OCO funding
** Reflects the FY 2011 President's Budget Request

Numbers may not add due to rounding

FY 2012 Program Acquisition Costs by Weapon System

P-8A Poseidon

The P-8A Poseidon is an all-weather, twin engine, commercial derivative of the Boeing 737 aircraft. This land-based, network enabled, maritime Patrol aircraft is designed to sustain and improve armed maritime and littoral Intelligence, Surveillance, and Reconnaissance (ISR) capabilities in traditional, joint, and combined roles to counter changing and emerging threats.

Mission: The P-8A Poseidon recapitalizes the Maritime Patrol Anti-Submarine Warfare (ASW), Anti-Surface Warfare (ASuW), and armed Intelligence, Surveillance and Reconnaissance (ISR) in maritime and littoral areas above, on and below the surface of the ocean. Provides the Joint warfighter lethality on datum – the only Defense platform fielding this operationally agile, tactically responsive capability.

FY 2012 Program: Funds 11 P-8A aircraft, associated spares, and provides the advance procurement resources for 13 FY 2013 aircraft, and continues the research, development, and testing of the P-8A systems. The aircraft procurements are tightly coupled to the P-3 retirement rates.

Prime Contractors: Airframe: The Boeing Company, Kent, WA
Engine: CFM International (General Electric Aviation and SNECMA), Cincinnati, OH

P-8A Poseidon

	FY 2010*		FY 2011**		FY 2012 Base Budget		FY 2012 OCO Budget		FY 2012 Total Request	
	$M	Qty	$M	Qty	$M	Qty	$M	Qty	$M	Qty
RDT&E	1,138.7	-	929.2	-	622.7	-			622.7	-
Procurement	1,797.4	6	1,990.6	7	2,275.4	11	-		2,275.4	11
Spares	104.7	-	72.4	-	98.3				98.3	-
	3,040.8	6	2,992.3	7	2,996.5	11	-	-	2,996.5	11

* FY 2010 & FY 2011 include Base and OCO funding
** Reflects the FY 2011 President's Budget Request

Numbers may not add due to rounding
No modification funding included

Command, Control, Communications, and Computer (C4) Systems

The Department is transforming and developing new concepts for the conduct of future joint military operations. The overarching goal is full spectrum dominance—defeat of any adversary or control of any situation across the full range of military operations—achieved through a broad array of capabilities enabled by an interconnected network of sensors, shooters, command, control, and intelligence. This network-based interconnectivity increases the operational effectiveness by assuring access to the best possible information by decision-makers at all levels, thus allowing dispersed forces to communicate, maneuver, share a common user-defined operating picture, and successfully complete assigned missions more efficiently. Net-centricity transforms the way that information is managed to accelerate decision making, improve joint warfighting, and create intelligence advantages. Hence, all information is visible, available, usable and trusted—when needed and where needed—to accelerate the decision cycles.

Net-centricity is a service-based architecture pattern for information sharing. It is being implemented by the Command, Control, Communications, Computer, and Intelligence (C4I) community via building joint architectures and roadmaps for integrating joint airborne networking capabilities with the evolving ground, maritime, and space networks. It encompasses the development of technologies like gateways, waveforms, network management, and information assurance.

FY 2012 Command, Control, Communications, and Computers (C4) Systems – Base and OCO: $10.9 Billion

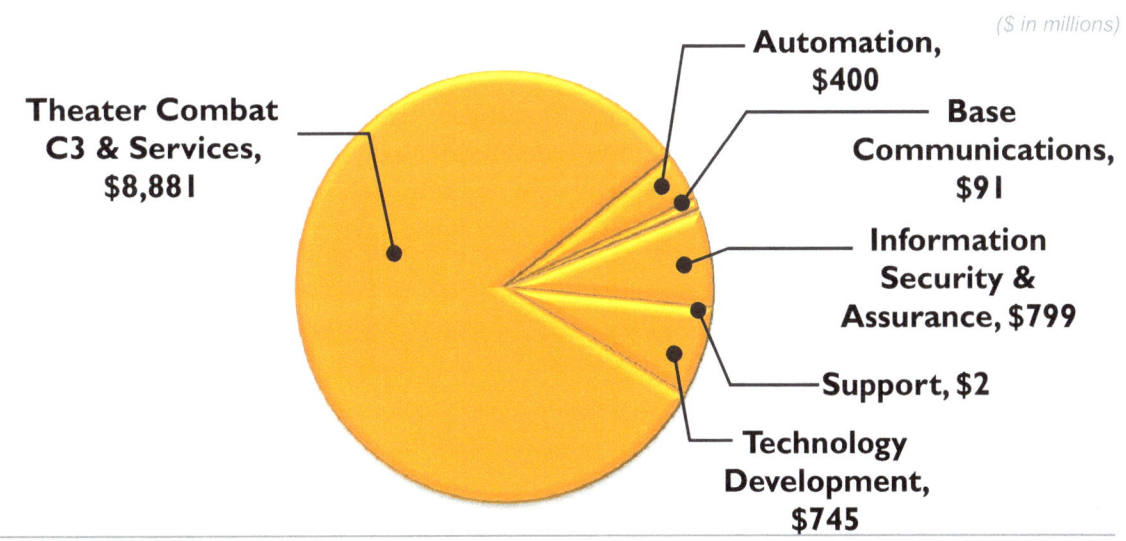

($ in millions)

- Theater Combat C3 & Services, $8,881
- Automation, $400
- Base Communications, $91
- Information Security & Assurance, $799
- Support, $2
- Technology Development, $745

Source: FY 2012 PRCP – Investment Categorization
Numbers may not add due to rounding

FY 2012 Program Acquisition Costs by Weapon System

Joint Tactical Radio System

DOD - JOINT

The Joint Tactical Radio System (JTRS) is a joint DoD effort to develop, produce, integrate, test, and field a family of software-defined, secure, multi-channel, digital radios that will be interoperable with existing radios and increase communication and networking capabilities for mobile and fixed sites. The program encompasses ground, airborne, vehicular, maritime, and small form factor variants of the radio hardware, 17 Increment 1 waveforms for porting into the JTRS hardware, and network management applications. All JTRS products are being developed in a joint environment to ensure interoperability and the enhancement of hardware and software commonality and reusability.

Mission: The JTRS products will simultaneously receive, transmit, and relay voice, data, and video communications with hardware configurable, software programmable, multi-band, and multi-mode network capable systems.

FY 2012 Program: Funds the design, development, and manufacture of JTRS engineering development models (EDMs) and low rate initial production (LRIP), to include hardware and software, as well as sustainment of fielded radios and certified waveforms.

Prime Contractors: The Boeing Company, Anaheim, CA
Lockheed Martin Corporation, Manassas, VA
ViaSat Incorporated, Carlsbad, CA
BAE Systems/Rockwell Collins Data Link Solutions, L.L.C., Cedar Rapids, IA
General Dynamics Decision Systems, Inc., Scottsdale, AZ
ITT Corporation, Fort Wayne, IN

Joint Tactical Radio System

	FY 2010*		FY 2011**		FY 2012					
					Base Budget		OCO Budget		Total Request	
	$M	Qty	$M	Qty	$M	Qty	$M	Qty	$M	Qty
RDT&E	857.5		687.7		688.1	-	-		688.1	-
Procurement	30.3	140	209.6	2,439	775.8	17,120	0.5	6	776.3	17,126
O&M	33.4		69.9		77.5	-	-		77.5	-
Spares	-	-	-	-	-	-	-	-	-	-
Total	921.2	140	967.2	2,439	1,541.4	17,120	0.5	6	1,541.9	17,126

FY 2010 & FY 2011 include Base and OCO funding
**Reflects the FY 2011 President's Budget Request*

Numbers may not add due to rounding

C4 SYSTEMS

FY 2012 Program Acquisition Costs by Weapon System

Early – Infantry Brigade Combat Team (E-IBCT) Modernization

The Army intends to transition the E-IBCT program of record (POR) to Capability Package based modernization. Following the acquisition of these LRIP quantities, the E-IBCT program will be completed and further SUGV acquisition delegated to the Army.

In place of place of E-IBCT program, the transition calls to continue development and sustainment of the right equipment in an incremental and iterative manner to ensure that Soldiers and units have the capabilities worthy of continuance will be successful across the full range of military operations today and into the future.

FY 2012 Program: Procure 1 additional brigade set of the Network Integration Kit (NIK) (quantity of 100) and two additional brigade sets of the Small Unmanned Ground Vehicle (SUGV) (an additional 78 units). The additional NIKs will include E-IBCT capable Ground Mobile Radios (GMR) radios as NIK subcomponents. The remaining elements of the E-IBCT program (Class 1 UAV, Tactical and Urban Unattended Ground Sensors (T/U-UGS)) are cancelled.

Prime Contractors: The Boeing Company, St. Louis, MO
Science Applications International Corporation (SAIC), Torrance, CA

Early - Infantry Brigade Combat Team (E-IBCT) Modernization

	FY 2010*		FY 2011**		FY 2012 Base Budget		FY 2012 OCO Budget		FY 2012 Total Request	
	$M	Qty	$M	Qty	$M	Qty	$M	Qty	$M	Qty
RDT&E	1,875	-	1,568	-	506	-	-	-	506	-
Procurement	211	-	683	-	243	-	-	-	243	-
Spares									-	-
Total	2,086	-	2,251	-	749	-	-	-	749	-

* FY 2010 & FY 2011 include Base and OCO funding
** Reflects the FY 2011 President's Budget Request

Numbers may not add due to rounding

Warfighter Information Network-Tactical

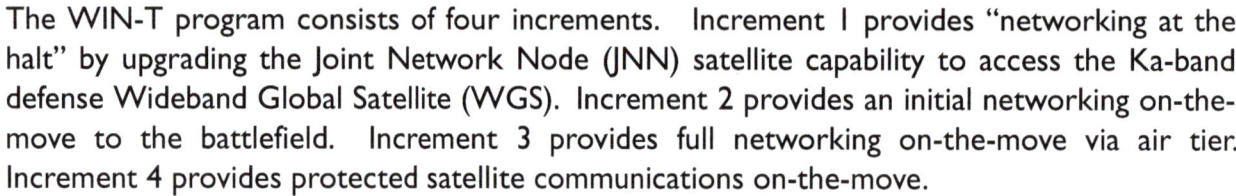

The Warfighter Information Network- Tactical (WIN-T) is the Army's on-the-move, high speed, high capability backbone communications network, linking Warfighters in the battlefield with the Global Information Grid (GIG). This network is intended to provide command, control, communications, computers, intelligence, surveillance and reconnaissance (C4ISR) support capabilities. The system is being developed as a network for reliable, secure and seamless video, data, imagery and voice services for the Warfighters in the theater to enable decisive combat actions.

The WIN-T program consists of four increments. Increment 1 provides "networking at the halt" by upgrading the Joint Network Node (JNN) satellite capability to access the Ka-band defense Wideband Global Satellite (WGS). Increment 2 provides an initial networking on-the-move to the battlefield. Increment 3 provides full networking on-the-move via air tier. Increment 4 provides protected satellite communications on-the-move.

Mission: The WIN-T program provides the United States Army with a transformational modernized network. Using satellite, air, and ground layers, it delivers the fully mobile, flexible, dynamic networking capability needed to support a highly dispersed force over a noncontiguous area.

FY 2012 Program: Procures and continues to field WIN-T Inc 1 to the Army, with a Ka satellite upgrade. WIN-T Inc 2 is currently in Limited Rate Initial Production (LRIP) in anticipation of its Initial Operational Test in FY 2012, and WIN-T Inc 3 continues in its Engineering, Manufacturing, and Development (EMD) phase to deliver the full networking on the move, including the airborne tier.

Prime Contractor: General Dynamics Corporation, Taunton, MA

Sub Contractor: Lockheed Martin Corporation, Gaithersburg, MD

Warfighter Information Network-Tactical

	FY 2010*		FY 2011**		FY 2012					
					Base Budget		OCO Budget		Total Request	
	$M	Qty	$M	Qty	$M	Qty	$M	Qty	$M	Qty
RDT&E	164.0		190.9		298.0				298.0	-
Procurement	610.6		421.8		974.2	3,931	0.5		974.7	-
Spares									-	-
Total	774.6	-	612.7	-	1,272.2	3,931	0.5	-	1,272.7	-

* FY 2010 & FY 2011 include Base and OCO funding
** Reflects the FY 2011 President's Budget Request

Numbers may not add due to rounding

FY 2012 Program Acquisition Costs by Weapon System

Ground Programs

The Department continues to modernize its ground force capabilities to ensure the United States remains a dominant force capable of operating in all environments across the full spectrum of conflict. The Army and Marine Corps equip each soldier and marine with the best equipment available to succeed in both today's and tomorrow's operations.

Modernization and upgrade of selected core systems is a continuous process. Some of the existing programs are targeted for upgrades to include howitzers, Stryker vehicles, M1 Abrams, Bradley Fighting Vehicle, and the Light Armored Vehicle (LAV).

The Army is focused on developing a Ground Combat Vehicle (GCV) to provide a new infantry fighting vehicle to the war fighter. The GCV has the design growth to adapt to capabilities as the operational environment changes and technology matures to position soldiers for long-term success. The Marine Corps is developing the Marine Personnel Carrier (MPC), an advanced generation armored personnel carrier that would provide general support lift to the marine infantry in the ground combat element based maneuver task force.

FY 2012 Ground Programs – Base and OCO: $16.1 Billion

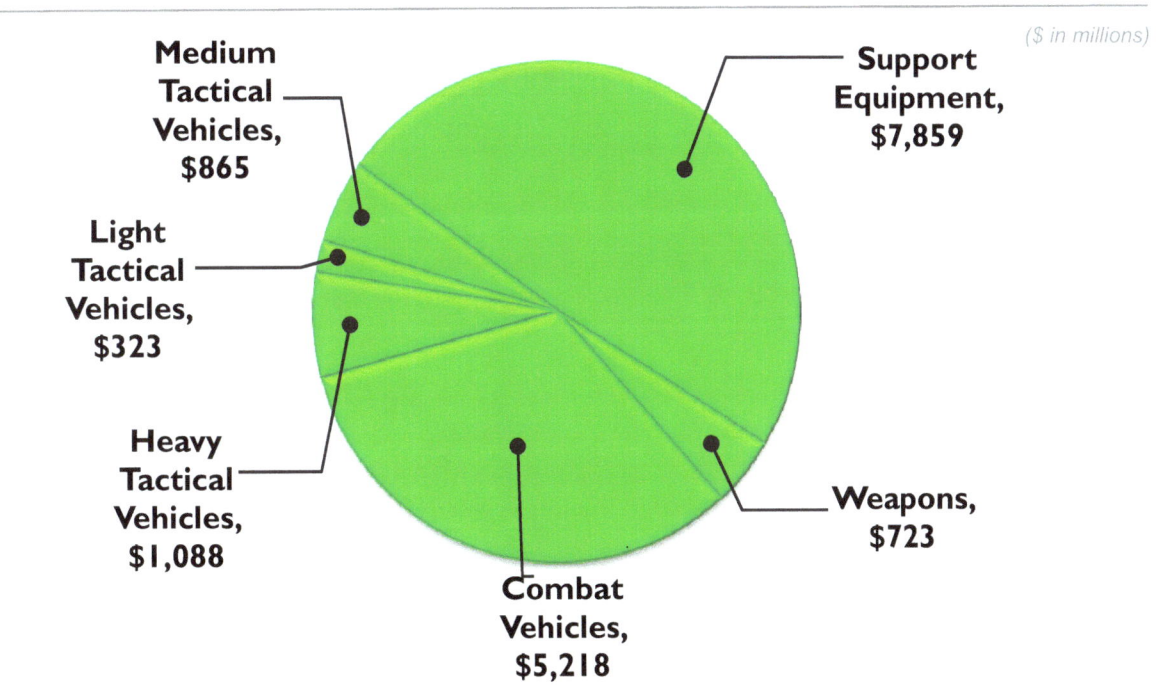

($ in millions)

- Medium Tactical Vehicles, $865
- Light Tactical Vehicles, $323
- Heavy Tactical Vehicles, $1,088
- Combat Vehicles, $5,218
- Weapons, $723
- Support Equipment, $7,859

Source: FY 2012 PRCP – Investment Categorization
Numbers may not add due to rounding

FY 2012 Program Acquisition Costs by Weapon System

Joint Light Tactical Vehicle

DoD - JOINT

The Joint Light Tactical Vehicle (JLTV) is a joint program currently in development for the Army and Marine Corps. The JLTV is intended to replace the High Mobility Multipurpose Wheeled Vehicle (HMMWV), which is the current light tactical vehicle. The JLTV concept is based on a family of vehicles focused on scalable armor protection and vehicle agility, and mobility required of the light tactical vehicle fleet. The JLTV will provide defensive measures to protect troops while in transport, increase payload capability, and achieve commonality of parts and components to reduce the overall life cycle cost of the vehicle. The JLTV project seeks to optimize performance, payload, and protection of the vehicle and crew while ensuring a design that is transportable by CH-47, CH-53, and C-130 aircraft.

Mission: As a light tactical vehicle, JLTV will be capable of performing multiple mission roles, and will be designed to provide protected, sustained, networked mobility for personnel and payloads across the full range of military operations. There are three mission role variants: General Purpose 3,500 lb; Infantry Carrier 4,500 lb; and Utility 5,100 lb.

FY 2012 Program: If approved in 4th quarter of FY 2011, the Army and Marine Corps will continue to develop the vehicle and an affordable manufacturing process.

Prime Contractor: Currently in Technology Development

Joint Light Tactical Vehicle

	FY 2010*		FY 2011**		FY 2012 Base Budget		FY 2012 OCO Budget		FY 2012 Total Request	
	$M	Qty	$M	Qty	$M	Qty	$M	Qty	$M	Qty
RDT&E USA	30.9	-	52.9	-	172.1	-	-	-	172.1	-
RDT&E USMC	53.0	-	31.8	-	71.8	-	-	-	71.8	-
Spares USA									-	-
Spares USMC									-	-
Total	83.9	-	84.7	-	243.9	-	-	-	243.9	-

* FY 2010 & FY 2011 include Base and OCO funding
** Reflects the FY 2011 President's Budget Request

Numbers may not add due to rounding

GROUND VEHICLES

Family of Heavy Tactical Vehicles

Army photo of a PLS

The Family of Heavy Tactical Vehicles (FHTV) consists of the Palletized Load System (PLS) and the Heavy Expanded Mobility Tactical Truck (HEMTT). The PLS entered service in 1993 and consists of a 16.5 ton, 10 wheel tactical truck with self load/unload capability. The PLS carry payload on flat rack cargo bed, trailer, or International Standards Organization (ISO) containers. The HEMTT is a 10-ton, 8 wheel (8x8) truck that comes in several configurations: The Tanker to refuel tactical vehicles and helicopters, Tractor to tow the Patriot missile system and Multi-Launch Rocket System (MLRS), Wrecker to recover vehicles, and Cargo truck with a materiel handling crane. The HEMTT entered service in 1982.

Mission: Provides transportation of heavy cargo to supply and re-supply combat vehicles and weapons systems. The PLS is fielded to transportation units, ammunition units, and to forward support battalions with the capability to self-load and transport a 20 ft. ISO container. The HEMTT A4 is an important truck to transport logistics behind quick-moving forces such as the M-1 Abrams and Stryker. The HEMTT is used in line haul, local haul, unit resupply, and other missions throughout the tactical environment to support modern and highly mobile combat units. The HEMTT trucks carry all types of cargo, especially ammunition and fuel.

FY 2012 Program: Procures various FHTV vehicles, trailers, and tracking systems to fill urgent theater requirements and modernize the Heavy Tactical Vehicle fleet for the Active, National Guard, and Reserve units.

Prime Contractor: Oshkosh Truck Corporation, Oshkosh, WI

Family of Heavy Tactical Vehicles

	FY 2010*		FY 2011**		FY 2012 Base Budget		FY 2012 OCO Budget		FY 2012 Total Request	
	$M	Qty	$M	Qty	$M	Qty	$M	Qty	$M	Qty
RDT&E	8.1	-	3.5	-	5.5	-	-	-	5.5	-
Procurement	1,402.6	1,490	738.4	1,173	627.3	1,569	47.2	29	674.5	1,598
Spares									-	-
Total	1,410.7	3,162	741.9	1,173	632.8	1,569	47.2	29	680.0	1,598

* FY 2010 & FY 2011 include Base and OCO funding
** Reflects the FY 2011 President's Budget Request

Numbers may not add due to rounding

FY 2012 Program Acquisition Costs by Weapon System

Family of Medium Tactical Vehicles

The Family of Medium Tactical Vehicles (FMTV) is a family of diesel powered trucks in the 2 1/2 ton and 5 ton payload class. The vehicle first went into service in 1996. It capitalizes on current state of the art automotive technology including a diesel engine, automatic transmission, and central tire inflation system (CTIS). The use of common chassis, engines, tires, and cabs are features over 80 percent commonality of parts between models and weight classes, which significantly reduces the logistics burden and operating costs. Numerous models perform a wide variety of missions including cargo transport (cargo model), vehicle recovery operations (wrecker), construction (dump), line haul (tractor), and airdrop missions, and civil disaster relief. The FMTV also serves as the platform for the High Mobility Artillery Rocket System (HIMARS) and support vehicle for the Patriot missile.

Mission: The FMTV provides unit mobility and resupply of equipment and personnel for rapidly deployable worldwide operations on primary and secondary roads, trails, cross-country terrain, and in all climatic conditions. It is strategically deployable in C-5, C-17, and C-130 aircraft. Experience in Iraq led to the development of an up-armored cab known as the Low Signature Armored Cab (LSAC) for installation on FMTV vehicles that adds ballistic and mine blast protection for the crew.

FY 2012 Program: Procures 2,390 Medium Tactical Vehicles in the baseline budget and 32 vehicles in the Overseas Contingency Operations budget to support the Army modular transformation effort to modernize the tactical wheeled vehicle fleet for medium size trucks.

Prime Contractor: Oshkosh Corporation

Family of Medium Tactical Vehicles

	FY 2010*		FY 2011**		FY 2012 Base Budget		FY 2012 OCO Budget		FY 2012 Total Request	
	$M	Qty	$M	Qty	$M	Qty	$M	Qty	$M	Qty
RDT&E	5.5	-	3.7	-	4.0	-	-	-	4.0	-
Procurement	1,344.3	8,637	1,434.5	4,652	432.9	2,390	11.1	32	444.0	2,422
Spares	-	-	-	-	-	-	-	-	-	-
Total	1,349.8	8,637	1,438.2	4,652	436.9	2,390	11.1	32	448.0	2,422

* FY 2010 & FY 2011 include Base and OCO funding
** Reflects the FY 2011 President's Budget Request

Numbers may not add due to rounding

FY 2012 Program Acquisition Costs by Weapon System

M-1 Abrams Tank Upgrade

US Army Photo

The M1 Abrams is the Army's main battle tank, which first entered service in 1980. It was produced from 1978 until 1992. Since then, the Army has modernized it with a series of upgrades to improve its capabilities. The current M1 Abrams tank modernization effort supports two variants. The M1A1 Situational Awareness (SA) and the M1A2 System Enhancement Program (SEP). The M1A1 SA modernization includes steel encased depleted uranium for increased frontal and turret side armor protection, suspension improvements, an advanced computer system with embedded diagnostics, a second generation thermal sensor, and a laser rangefinder to designate targets from increased distances. The M1A2 SEP tank modernization includes a commander's independent thermal weapons station, position navigation equipment, improved fire control system, and an improved AGT1500 turbine engine.

Mission: The M1A2 Abrams is the Army's main battle tank that provides mobile and protected firepower for battlefield superiority against heavy armor forces.

FY 2012 Program: Upgrades and fields M1A2 SEP tanks to armor units including the 1st Armor Division.

Prime Contractor: General Dynamics Corporation, Sterling Heights, MI

M-1 Abrams Tank Upgrade

| | FY 2010* | | FY 2011** | | FY 2012 | | | | | |
| | | | | | Base Budget | | OCO Budget | | Total Request | |
	$M	Qty	$M	Qty	$M	Qty	$M	Qty	$M	Qty
RDT&E	93.8	-	107.5	-	9.7	-	-	-	9.7	-
Procurement	185.0	22	183.0	21	181.3	21			181.3	21
Spares									-	-
Total	278.8	22	290.5	21	191.0	21	-	-	191.0	21

* FY 2010 & FY 2011 include Base and OCO funding
** Reflects the FY 2011 President's Budget Request

Numbers may not add due to rounding

FY 2012 Program Acquisition Costs by Weapon System

Stryker Family of Armored Vehicles

US Army Photo

Stryker is a 19-ton wheeled armored vehicle that will provide the Army a family of ten different vehicles. The Stryker can be deployed by C-130, C-17, and C-5 aircraft and be combat-capable upon arrival in any contingency area. It can reach speeds of 62 mph on the highway and has a maximum range of 312 miles.

There are two basic versions, which include the Infantry Carrier Vehicle (ICV) and the Mobile Gun System (MGS). There are eight different configurations, which include the Reconnaissance Vehicle (RV); Anti-Tank Guided Missile (ATGM); Nuclear, Biological, Chemical, and Radiological Vehicle (NBCRV); Medical Evacuation Vehicle (MEV) Commander's Vehicle (CV); Fire Support Vehicle (FSV); Mortar Carrier (MC); and Engineer Squad Vehicle (ESV).

Mission: The Stryker vehicle is designed to enable the Brigade Combat Team to maneuver more easily in close and urban terrain while providing protection in open terrain. It fills the Army's current transformation goal to equip a strategically deployable brigade using a C-17 or C-5 and operationally deployable brigade using a C-130 that is capable of rapid movement anywhere on the globe in a combat ready configuration.

FY 2012 Program: Procures 100 Stryker vehicles in FY 2012

Prime Contractor: General Dynamics Corporation, Sterling Heights, MI

Stryker Family of Armored Vehicles

	FY 2010*		FY 2011**		FY 2012					
					Base Budget		OCO Budget		Total Request	
	$M	Qty	$M	Qty	$M	Qty	$M	Qty	$M	Qty
RDT&E	96.3	-	136.3	-	101.4	-			101.4	-
Procurement	512.8	93	299.5	83	633.0	100			633.0	100
Spares					99.6				99.6	-
Total	609.1	93	435.8	83	834.0	100	-	-	834.0	100

* FY 2010 & FY 2011 include Base and OCO funding
** Reflects the FY 2011 President's Budget Request

Numbers may not add due to rounding

Missile Defense

Missile Defense is a general term for air and missile defense. This category includes cruise missile, air and ballistic missile defense systems program development. The Missile Defense Agency and the Army are the program developer's. Missile Defense includes all components designed to defeat hostile ballistic missiles of various ranges. A missile defense system includes interceptor missiles, as well as the associated sensors and command, control, battle management, and communications. Other significant investments include construction, targets and countermeasures, and research, development, testing, and evaluation activities. Encompassed in this category are all programs that are either critical to the functionality of missile defense or support missile defense as a primary mission.

The Department continues to invest and build inventories of air and missile defense capabilities, such as the Patriot Advanced Capability (PAC-3) missiles, Standard Missile-3 (SM-3) interceptors, Terminal High Altitude Area Defense interceptors (THAAD), and the Army Navy/Transportable Radar Surveillance – Model 2 (AN/TPY-2). The Department continues to seek expanded international efforts for missile defense with allies and friends to provide pragmatic and cost-effective missile defense capabilities.

FY 2012 Missile Defense – Base and OCO: $10.1 Billion

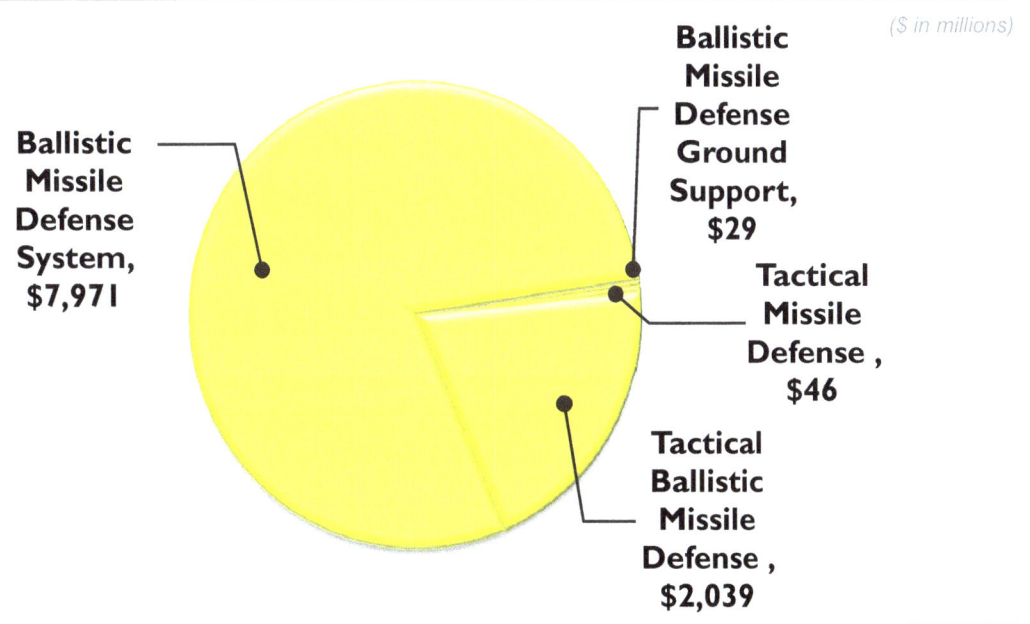

($ in millions)

- Ballistic Missile Defense System, $7,971
- Ballistic Missile Defense Ground Support, $29
- Tactical Missile Defense, $46
- Tactical Ballistic Missile Defense, $2,039

Numbers may not add due to rounding

Source: FY 2012 PRCP – Investment Categorization
Note: Total resource request of $10.1 billion does not include the Missile Defense Agency's Science and Technology $0.4 billion, the $0.07 billion of Military Construction nor the $0.02 billion of Operation and Maintenance, Defense Wide. Due to rounding these additives appropriations do not add to the Ballistic Missile Defense $10.7 billion.

FY 2012 Program Acquisition Costs by Weapon System

Ballistic Missile Defense

Ballistic Missile Defense (Army and Joint Staff)

| | FY 2010* | | FY 2011** | | FY 2012 | | | | | |
| | | | | | Base Budget | | OCO Budget | | Total Request | |
	$M	Qty	$M	Qty	$M	Qty	$M	Qty	$M	Qty
RDT&E, Army										
PATRIOT/PAC-3 [1]	11.0	-	11.5	-	44.3	-	-	-	44.3	-
PAC-3/MSE [2/3]	-	-	62.5	-	89.0	-	-	-	89.0	-
PATRIOT/MEADS [4]	571.0	-	467.1	-	406.6	-	-	-	406.6	-
JLENS [5]	317.1	-	372.5	-	344.7	-	-	-	344.7	-
AIAMD [6/7]	164.7	-	251.1	-	270.6	-	-	-	270.6	-
Subtotal, RDT&E	1,064.3	-	1,164.7	-	1,155.2	-	-	-	1,155.2	-
Procurement, Army [8]										
Patriot/PAC-3	396.4	59	544.2	78	735.8	88	-	-	735.8	88
PAC-3/MSE [1/2]	-	-	-	-	75.0	-	-	-	75.0	-
Subtotal, Proc	396.4	59	544.2	78	810.8	88	-	-	810.8	88
RDT&E, DW										
JIAMDO [9]	97.0	-	94.6	-	79.9	-	-	-	79.9	-
Subtotal Non-MDA	1,557.7	-	1,803.5	-	2,045.9	-	-	-	2,045.9	-
Total Ballistic Missile Defense	9,449.6	-	10,219.9	-	10,671.6	-	-	-	10,671.6	-

[1] Patriot Advanced Capability-3 (PAC-3)

[2] Patriot Advanced Capability-3 (PAC-3)/Missile Segment Enhancement (MSE)

[3] PAC-3/MSE was executed in Patriot/MEADS CAP in FY 2010

[4] Medium Extended Air Defense Systems (MEADS)

[5] Joint Land Attack Cruise Missile Defense Elevated Netted Sensor System (JLENS)

[6] Army Integrated Air and Missile Defense (AIAMD)

[7] AIAMD was added to this papers content in President's Budget Request FY 2012

[8] Modifications and spares resources are included.

[9] Joint Integrated Air and Misile Defense Organization (JIAMDO)

* FY 2010 & FY 2011 include Base and OCO funding
** Reflects the FY 2011 President's Budget Request

Numbers may not add due to rounding

FY 2012 Program Acquisition Costs by Weapon System

Ballistic Missile Defense

The Ballistic Missile Defense (BMD) programs provide defense capabilities to the territory of the United States against ballistic missile threats from rogue nations and accidental or unauthorized launches. Additionally, ballistic missile defense will defend our U.S. deployed forces, Allies, and friends against regional threats. Ballistic missile defense is managed as an integrated layered system to include researching concepts, developing and fielding the earliest possible capability in sea, ground, space and air to intercept any range of threat in the boost, midcourse or terminal phases of flight trajectory. Major elements include Patriot Advanced Capability-3 (PAC-3), AEGIS BMD, Terminal High Altitude Area Defense (THAAD), Patriot Medium Extended Air Defense System (MEADS), PAC-3 Missile Segment Enhancement (MSE), Ground-based Midcourse Defense (GMD), and the Joint Land Attack Cruise Missile Defense Elevated Netted Sensor System (JLENS) components, all of which support the Phased Adaptive Approach; a Presidential initiative.

Mission: Develop, field, and sustain a missile defense capability to defend the United States, its Allies, and U.S. deployed forces against rogue nation attacks, to close gaps and improve this capability against future threats, and to further develop options to defeat near-term and emerging threats.

FY 2012 Program: Continues the research, development, testing, fielding and conversion and integration of AEGIS BMD capable ships, along with the sustainment of ballistic missile defense programs.

Prime Contractors: Boeing, Lockheed Martin, Northrop Grumman, Raytheon

Ballistic Missile Defense

	FY 2010*		FY 2011**		FY 2012					
					Base Budget		OCO Budget		Total Request	
	$M	Qty	$M	Qty	$M	Qty	$M	Qty	$M	Qty
RDT&E	8,032.1	-	8,713.9	-	7,812.1	-	-	-	7,812.1	-
Procurement	1,232.2	-	1,497.3	-	2,589.6	-	-	-	2,589.6	-
MILCON	98.7	-	-	-	67.2	-	-	-	67.2	-
BRAC	86.6	-	8.7	-	-	-	-	-	-	-
O&M	-	-	-	-	202.7	-	-	-	202.7	-
Spares	-	-	-	-	-	-	-	-	-	-
Total	9,449.6	-	10,219.9	-	10,671.6	-	-	-	10,671.6	-

Note: Funding includes more than Investment resources. Numbers may not add due to rounding

Includes the Missile Defense Agency's - Base Realignment and Closure, Military Construction, Science and Technology, and Operation and Maintenance resources.

Includes the Army's modifications and spare resources.

* FY 2010 & FY 2011 include Base and OCO funding

** Reflects the FY 2011 President's Budget Request

FY 2012 Program Acquisition Costs by Weapon System

Aegis Ballistic Missile Defense

The Aegis Ballistic Missile Defense System (BMDS), is a key sea-based element of the Missile Defense Agency's (MDA's) program, and is building upon the existing U.S. Navy Aegis Weapons System (AWS) and Standard Missile (SM) infrastructures. Aegis provides a forward-deployable, mobile capability to detect and track Ballistic Missiles of all ranges, and the ability to destroy Short- Medium-, Intermediate-Range Ballistic Missile, and selected long-range class threats in the midcourse phase of flight. Spiral upgrades to both the Aegis BMD Weapon System (AWS) and the SM-3 configurations enables Aegis BMD to provide effective, supportable defensive capability against more difficult threats, including Long-Range Ballistic Missiles, and expand the capability to counter limited engagements in the terminal phase of flight.

Mission: The Aegis BMD is delivering an enduring, operationally effective and supportable BMDS capability on Aegis cruisers and destroyers to defend the nation, deployed forces, friends, and allies and to incrementally increase this capability by delivering evolutionary incremental improvements as part of the BMDS upgrades.

FY 2012 Program: Continues procuring the Aegis Weapon System upgrades for five additional Aegis ships. Completes manufacturing of 30 SM-3 Block IB interceptors incrementally funded with RDT&E resources. Completes development of Aegis BMD 4.0.1 and SM-3 Block IB. Continues the development of the Aegis BMD 5.0 and 5.1. Provides for the initial production of 46 SM-3 Block IB missiles.

Prime Contractors: Aegis Weapon System: Lockheed Martin Corporation, Moorestown, NJ
SM-3 Interceptor: Raytheon Company, Tucson, AZ

AEGIS Ballistic Missile Defense

	FY 2010*		FY 2011**		FY 2012 Base Budget		FY 2012 OCO Budget		FY 2012 Total Request	
	$M	Qty[1]	$M	Qty[1]	$M	Qty[1]	$M	Qty[1]	$M	Qty[1]
RDT&E	1,419.0	-	1,467.3	-	960.3	-	-	-	960.3	-
Procurement	225.6	6	94.1	8	565.4	46	-	-	565.4	46
Spares	-	-	-	-	-	-	-	-	-	-
Total	1,644.6	6	1,561.4	8	1,525.7	46	-	-	1,525.7	46

[1] Quantity is associated with SM-3 interceptors
* FY 2010 & FY 2011 include Base and OCO funding
** Reflects the FY 2011 President's Budget Request

Numbers may not add due to rounding

FY 2012 Program Acquisition Costs by Weapon System

THAAD Ballistic Missile Defense — DOD - JOINT

The Terminal High Altitude Area Defense (THAAD) is a Missile Defense Agency (MDA) Program and a key element of the Ballistic Missile Defense System (BMDS). The THAAD Tactical Groups will provide rapidly-transportable interceptors, using "Hit-To-Kill" technology to destroy ballistic missiles inside and outside the Atmosphere. A Battery consists of 6 truck-mounted launchers, 48 interceptors (8 per launcher), 1 AN/TPY-2 radar, and 1 tactical fire control/communications (TFCC) component.

Mission: Provide any Combatant Commander with the rapidly deployable, ground-base missile defense components that deepen, extend and compliment the BMDS, which will defeat ballistic missiles of all types and ranges in all phases of flight.

FY 2012 Program: Continues the development, testing, integration, fielding and sustainment of the THAAD components. Completes the initial fielding to the Army of two Batteries at Fort Bliss, TX. Continues an extensive training program for the soldiers on the use and the maintenance of the components as an operational unit. Continues the planning, development, and analysis of THAAD Launch on Network, which would provide the ability to initiate a THAAD engagement based on information provided by ballistic missile defense sensors outside of the launching THAAD Battery. Continues the purchase of Batteries 3 and 4, and adds the purchase of Battery 5. Completes the manufacturing of 50 Interceptors incrementally funded with RDT&E resources. Increases THAAD missile manufacturing capability from 4 to 6 per month. Conducts flight testing at Kwajalein Atoll to allow engagement of longer range targets.

Prime Contractor: Lockheed Martin Corporation, Sunnyvale, CA

Terminal High Altitude Area Defense (THAAD)

	FY 2010*		FY 2011**		FY 2012					
					Base Budget		OCO Budget		Total Request	
	$M	Qty[1]	$M	Qty[1]	$M	Qty[1]	$M	Qty	$M	Qty[1]
RDT&E	690.1	-	436.5	-	290.5	-	-	-	290.5	-
Procurement	419.0	26	858.9	67	833.2	68	-	-	833.2	68
O&M	-	-	-	-	51.0	-	-	-	51.0	-
Total	1,109.1	26.0	1,295.4	67	1,174.7	68	-	-	1,174.7	68

[1] Quantity is associated with THAAD Interceptors
* FY 2010 & FY 2011 include Base and OCO funding
** Reflects the FY 2011 President's Budget Request

Numbers may not add due to rounding

MISSILE DEFENSE SYSTEMS

Patriot/PAC-3

The Army's PATRIOT Advanced Capability (PAC-3) missile is the latest improvement to the PATRIOT air and missile defense system. The PATRIOT is the only combat-proven system capable of defeating Tactical Ballistic Missiles (TBMs), cruise missiles, and air breathing threats worldwide. The combatant commanders demand additional PATRIOT capability to defeat growing threats to U.S. forces deployed in Overseas Contingency Operations. The Army will add two additional PATRIOT PAC-3 configuration battalions in FY 2011 and FY 2012 as part of the Grow-the-Army initiative. The Army and the Missile Defense Agency jointly continue to evolve the successful integration of PAC-3 capabilities into the Ballistic Missile Defense System (BMDS).

Mission: The PATRIOT system contributes to the BMDS overall situational awareness for short range terminal ballistic missile threats. It can cue other systems while protecting BMDS assets. The PATRIOT system is further enhanced by networked BMDS remote sensors supplying early warning data, thus, increasing the probability of successful threat engagement. The PAC-3 units are the combatant commanders most capable system to protect soldiers, Allies, and assets against these threats.

FY 2012 Program: Continues the procurement of 88 PAC-3 missiles and 36 Electronic Launcher Enhanced Systems (ELES) launchers capable of firing the PAC-3 missile. Provides for the testing and procurement of the latest PATRIOT system software upgrades, which will enhance PATRIOT capability against the current threat and continue to decrease fratricide risk.

Prime Contractor: Lockheed Martin Missiles and Fire Control, Dallas, TX

	FY 2010*		FY 2011**		FY 2012					
					Base		OCO		Total	
	$M	Qty[1]	$M	Qty[1]	$M	Qty[1]	$M	Qty[1]	$M	Qty[1]
RDT&E	11.0	-	11.5	-	44.3	-	-	-	44.3	-
Procurement[2]	341.3	59	480.2	78	662.2	88	-	-	662.2	88
Spares	10.5	-	7.0	-	6.7	-	-	-	6.7	-
Total	362.7	59	498.7	78	713.2	88	-	-	713.2	88

[1] Quantity is associated with PAC-3 Missiles
[2] No modification funding included
* FY 2010 & FY 2011 include Base and OCO funding
** Reflects the FY 2011 President's Budget Request

Numbers may not add due to rounding

PATRIOT/MEADS — DOD - JOINT

The Medium Extended Air Defense System (MEADS) is a cooperative effort among the United States, Germany, and Italy to develop a ground based air and tactical ballistic missile defense capability system as a replacement for PATRIOT (U.S. and Germany), Hawk (Germany), and Nike Hercules (Italy). The MEADS will be a highly mobile, tactically deployable system providing defense to critical assets from ballistic missiles, cruise missiles, and other air breathing threats. Mounted on wheeled vehicles, the system will include launchers carrying several interceptors along with advanced radars that will provide 360-degree coverage on the battlefield. Interceptors will use the latest hit-to-kill technology (directly hitting the target to destroy it). The cooperative effort will help promote interoperability within North Atlantic Treaty Organization forces and will help bridge the gap between short-range maneuver air and missile defense systems and the long-range Ballistic Missile Defense System elements. The Missile Segment Enhancement (MSE) is the primary missile for the system, which performs at an extended range beyond the PATRIOT Advanced Capability-3 missile.

Mission: The MEADS provides joint and coalition forces with critical assets to defend area protection against multiple and simultaneous attacks and with the capability to counter, defeat, or destroy the missiles or aerial vehicles. Further, MEADS will provide significant improvements in strategic deployability, transportability, mobility, and maneuverability.

FY 2012 Program: Continues the cooperative MEADS Memorandum of Understanding with Germany and Italy for the System Development and Demonstration (SDD) phase of the program by continuing the design and development of the system.

Prime Contractor: MEADS International, Orlando, FL

PATRIOT/MEADS

	FY 2010*		FY 2011**		FY 2012 Base Budget		FY 2012 OCO Budget		FY 2012 Total Request	
	$M	Qty	$M	Qty	$M	Qty	$M	Qty	$M	Qty
RDT&E	571.0	-	467.1	-	406.6	-	-	-	406.6	-
Procurement		-		-		-		-		-
Total	571.0	-	467.1	-	406.6	-	-	-	406.6	-

* FY 2010 & FY 2011 include Base and OCO funding
** Reflects the FY 2011 President's Budget Request

Numbers may not add due to rounding

FY 2012 Program Acquisition Costs by Weapon System

PAC-3/MSE Missile

DOD - JOINT

The Missile Segment Enhancement (MSE) missile evolves from the existing PATRIOT Advanced Capability (PAC-3) missile. It is a substantial performance improvement to the PAC-3. The MSE upgrade enhances the current PAC-3 missile design and improves it with a higher performance envelope, a dual pulse, 11-inch diameter Solid Rocket Motor (SRM) design, improved lethality, thermally hardened front-end, upgraded batteries, enlarged fixed fins, more responsive control surfaces, and upgraded guidance software. These improvements result in a more agile, lethal interceptor missile with enhanced Insensitive Munitions (IM) compliance. The MSE will meet U. S. operational requirements to include firing from the PATRIOT system, and is the internationally accepted missile for the Medium Extended Air Defense System (MEADS).

Mission: The MSE missile is a hit-to-kill, surface-to-air missile. Like the PAC-3, the missile provides the range, accuracy, and lethality to be effectively used against Tactical Ballistic Missiles (TBMs) that have chemical, biological, radiological, nuclear and conventional high explosive warheads. The MSE is capable of intercepting Tactical Ballistic Missiles, Cruise Missiles, and Air-Breathing threats. The MSE's expanded engagement envelope, dual pulse motor, and other upgrades mean it has a range extended beyond the existing PAC-3 missile, filling a critical performance gap. The MSE's higher probability of kill results in greater protection for the U. S. warfighters, coalition forces, and critical assets.

FY 2012 Program: Continues the development, testing and integration of the MSE into the PATRIOT system, along with limited development and testing for ongoing Patriot ground equipment upgrades.

Prime Contractor: Lockheed Martin Missiles and Fire Control, Dallas, TX

PAC-3/MSE

| | FY 2010* | | FY 2011** | | FY 2012 | | | | | |
| | | | | | Base Budget | | OCO Budget | | Total Request | |
	$M	Qty	$M	Qty	$M	Qty	$M	Qty	$M	Qty
RDT&E	-	-	62.5	-	89.0	-	-	-	89.0	-
Procurement	-	-	-	-	75.0	-	-	-	75.0	-
Total	-	-	62.5	-	163.9	-	-	-	163.9	-

* FY 2010 & FY 2011 include Base and OCO funding
** Reflects the FY 2011 President's Budget Request

Numbers may not add due to rounding

MISSILE DEFENSE SYSTEMS

Ground-based Midcourse Defense

DoD - JOINT

DoD Missile Defense Agency Photo

The Ground-based Midcourse Defense (GMD) element is a Missile Defense Agency program and a key component of the Ballistic Missile Defense System (BMDS), providing Combatant Commanders capability to engage ballistic missiles in the midcourse phase of flight. This phase, compared to boost or terminal, allows significant time for sensor viewing from multiple platforms and, thus, provides multiple engagement opportunities for hit-to-kill interceptors. The Ground-Based Interceptor (GBI) is made up of a three-stage, solid fuel booster and an exo-atmospheric kill vehicle. When launched, the booster missile carries the kill vehicle toward the target's predicted location in space. Once released from the booster, the 152 pound kill vehicle uses data received in-flight from ground-based radars and its own on-board sensors to hit directly the incoming missile by ramming the warhead with a closing speed of approximately 15,000 miles per hour. Interceptors are currently emplaced at Fort Greely, AK and Vandenberg AFB, CA. The GMD fire control centers have been established in Colorado and Alaska.

Mission: The GMD provides the Combatant Commanders capability to defend the United States, including Hawaii and Alaska, against long range ballistic missiles during the midcourse phase of flight.

FY 2012 Program: Continues the development and sustainment of the GMD weapon system, which includes the deployment of 26 GBIs at Fort Greely, AK, and 4 GBIs at Vandenberg AFB, CA. Provides for the continued use of the flight test rotation plan. Where older fielded Ground Based Interceptors will be configured for flight testing to support the Integrated Master Test Plan (IMTP) requirements. Continues the Stockpile Reliability Program (SRP) and component aging testing in order to understand the health of the deployed assets. Completes Missile Field 2 in a 14-silo configurations by the end of 2011.

Prime Contractor: Boeing Defense and Space (BDS), St. Louis, MO

Ground-based Midcourse Defense

	FY 2010*		FY 2011**		FY 2012					
					Base Budget		OCO Budget		Total Request	
	$M	Qty	$M	Qty	$M	Qty	$M	Qty	$M	Qty
RDT&E	1,022.0	-	1,346.2	-	1,161.0	-	-	-	1,161.0	-
Procurement	-	-	-	-	-	-	-	-	-	-
Total	1,022.0	-	1,346.2	-	1,161.0	-	-	-	1,161.0	-

* FY 2010 & FY 2011 include Base and OCO funding

** Reflects the FY 2011 President's Budget Request

Numbers may not add due to rounding

JLENS

DOD - JOINT

The Joint Land Attack Cruise Missile Defense Elevated Netted Sensor System (JLENS) is a critical part of the Army's future Integrated Air and Missile Defense (IAMD) architecture and Is a Joint Service interest program. A JLENS Orbit is comprised of two systems: a Fire Control Radar system and a Surveillance Radar system. Each system is comprised of a 74-meter tethered aerostat, a Mobile Mooring Station (MMS), a Communications and Processing Group (CPG), and associated Ground Support Equipment (GSE). The JLENS can stay aloft up to 30 days providing 24-hour radar coverage of the assigned area enabling the only elevated, persistent, long-range surveillance and fire control sensor capability. The JLENS first flight demonstration was successfully conducted on August 25, 2009.

Mission: The JLENS provides elevated, persistent, Over-The-Horizon (OTH) surveillance using its advanced sensor and networking technologies to provide 360-degree wide-area surveillance. The sectored precision tracking provides enabling protection of the U. S. forces, allies, and coalition forces, as well as critical geo-political assets primarily from cruise missiles, aircraft, and Unmanned Aerial Vehicles; secondarily from Tactical Ballistic Missiles and Large Caliber Rockets in the boost-phase; and tertiary situational awareness of Surface Moving Targets.

FY 2012 Program: Provides for the continued development, testing and integration of the JLENS program. Conducts the integration and testing of the JLENS Engineering, Manufacturing and Development (EMD) Orbit to meet the next key decision point in the fourth quarter of FY 2012. This will enable the first EMD Orbit to equip the first unit in the fourth quarter of FY 2013. The Army programmed 5 Orbits (1 EMD and 4 Procurement) between FY 2013-2017.

Prime Contractor: Raytheon, Andover, MA

Joint Land Attack Cruise Missile Defense Elevated Netted System

| | FY 2010* | | FY 2011** | | FY 2012 | | | | | |
| | | | | | Base Budget | | OCO Budget | | Total Request | |
	$M	Qty	$M	Qty	$M	Qty	$M	Qty	$M	Qty
RDT&E	317.1	-	372.5	-	344.7	-	-	-	344.7	-
Procurement	-	-	-	-	-	-	-	-	-	-
Total	317.1	-	372.5	-	344.7	-	-	-	344.7	-

* FY 2010 & FY 2011 include Base and OCO funding

** Reflects the FY 2011 President's Budget Request

Numbers may not add due to rounding

FY 2012 Program Acquisition Costs by Weapon System

Phased Adaptive Approach — DOD - JOINT

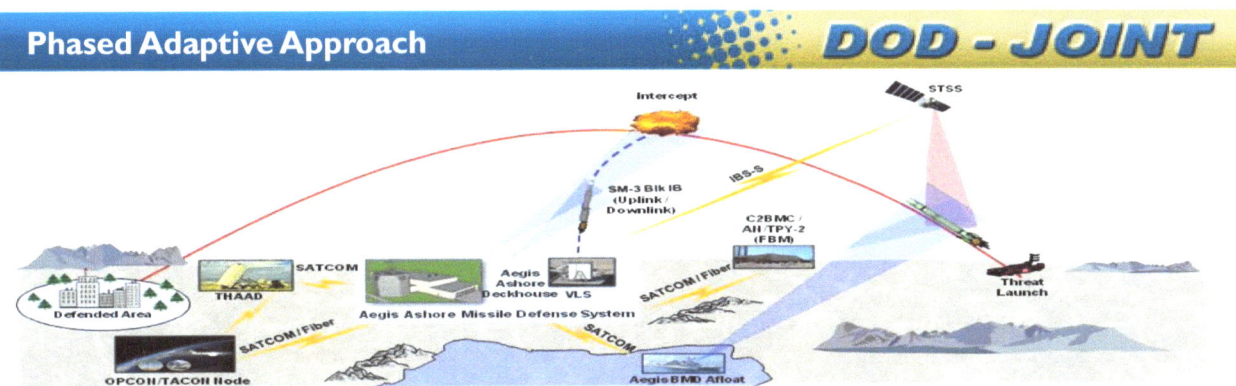

The Phased Adaptive Approach (PAA) is a response to the increased development and proliferation of short and medium range ballistic missiles in Iran and Around the world. The initial application of this approach addresses the threat posed to U.S. Allies and partners, as well as to U.S. deployed personnel in the Middle East and in Europe. By leveraging recent advances in sensor and missile technologies, the United States will aggressively counter this growing Regional threat with a more flexible and agile systems approach.

Mission: Starting in Europe, the United States is pursuing a four phased approach which will provide a more effective missile defense capability for defense of NATO territories and enhance U.S. homeland defense. The PAA will be complementary and interoperable with those being developed by NATO. Also, the PAA is applicable in other theaters around the world, and will be more adaptable and flexible to counter threat advances and provide increased defended areas over time. Aegis Ashore can adapt to the threat and be deployed/redeployed to areas needed to provide persistent coverage for the Geographic Combatant Commander.

FY 2012 Program: Continues the development and integration of the Aegis Ashore Missile Defense System. MDA will begin to install the Aegis Ashore Missile Defense Test Complex (AAMDTC) at Pacific Missile Range Facility (PMRF) on Kauai, Hawaii to provide proof of concept, system verification and validation of the first shore-based operation, support deployment decisions and upgrades of future incremental capabilities. Begins to procure long lead material for Host Nation (HN) #1.

Prime Contractor:
Aegis Weapon System: Lockheed Martin Corporation, Moorestown, NJ
SM-3 Interceptor: Raytheon Company, Tucson, AZ

Phased Adaptive Approach (PAA)

	FY 2010*		FY 2011**		FY 2012 Base Budget		FY 2012 OCO Budget		FY 2012 Total Request	
	$M	Qty	$M	Qty	$M	Qty	$M	Qty	$M	Qty
RDT&E	50.0	-	441.7	-	358.7	-	-	-	358.7	-
Procurement	-	-	-	-	261.4	-	-	-	261.4	-
MILCON	68.5	-	-	-	8.4	-	-	-	8.4	-
Total	118.5	-	441.7	-	628.4	-	-	-	628.4	-

* FY 2010 & FY 2011 include Base and OCO funding
** Reflects the FY 2011 President's Budget Request

Numbers may not add due to rounding

FY 2012 Program Acquisition Costs by Weapon System

Munitions and Missiles

Munitions is a general term for ammunition and missiles including conventional ammunition, bombs, missiles, warheads, and mines. This category includes conventional and nuclear weapons and weapons used for both tactical and strategic purposes. Many of the missiles and munitions are precision guided with the technical sophistication to allow guidance corrections during flight-to-target. Some programs include non-explosive articles that enhance the performance of other munitions. For example, the Joint Direct Attack Munitions (JDAM) adds guidance capability when attached to a gravity bomb, making it a "smart" bomb. *Note: Interceptor missiles supporting the missile defense mission are included in the Missile Defense section.*

The Department continues to build inventories of standoff weaponry, such as the Joint Air-to-Surface Standoff Missile, the Joint Standoff Weapon, and the Small Diameter Bomb.

FY 2012 Munitions and Missiles – Base and OCO: $11.0 Billion

($ in millions)

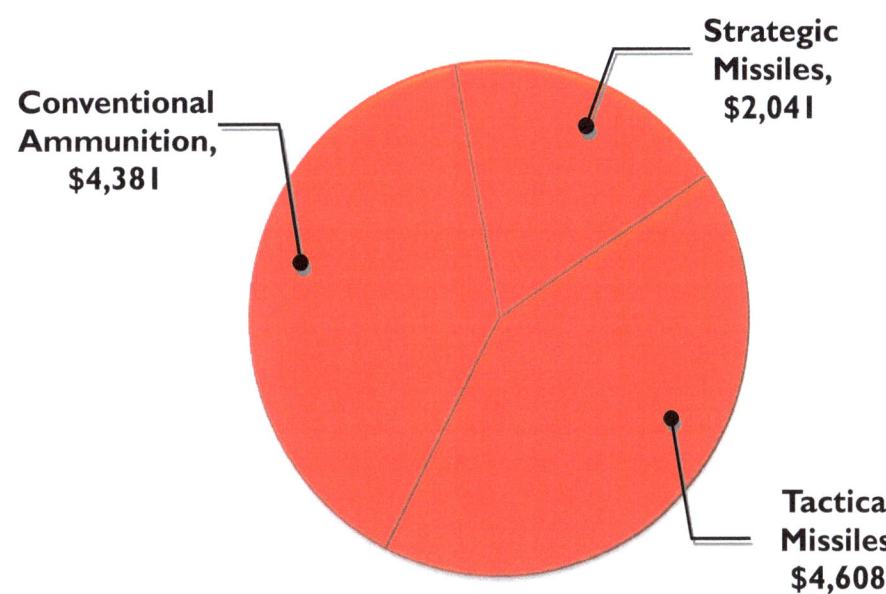

Source: FY 2012 PRCP – Investment Categorization
Numbers may not add due to rounding

FY 2012 Program Acquisition Costs by Weapon System

Advanced Med. Range Air-to-Air Missile

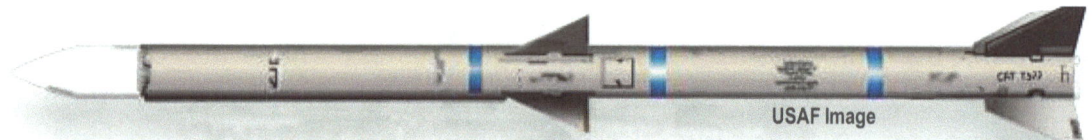
USAF Image

The Advanced Medium Range Air-to-Air Missile (AMRAAM) is an all-weather, all-environment radar guided missile developed to improve capabilities against very low-altitude and high-altitude, high-speed targets in an electronic countermeasures environment. The AMRAAM is a joint Navy/Air Force program led by the Air Force.

Mission: The mission of the AMRAAM is to destroy low and high altitude, high-speed enemy targets in an electronic countermeasures environment. AMRAAM is a fire-and-forget air-to-air missile, and has replaced the AIM-7 Sparrow as the U.S. military's standard BVR (Beyond Visual Range) intercept missile. The missile has undergone various service life improvements. The current generation AIM-120D, has a two-way data link, GPS-enhanced IMU, an expanded no-escape envelope, improved High-Angle Off-Boresight capability, and a 50% increase in range

FY 2012 Program: Continues full rate production as well as product improvements such as fuzing, guidance, and kinematics.

Prime Contractor: Raytheon Company, Tucson, AZ

Advanced Med. Range Air-to-Air Missile

	FY 2010*		FY 2011**		FY 2012					
					Base Budget		OCO Budget		Total Request	
	$M	Qty	$M	Qty	$M	Qty	$M	Qty	$M	Qty
RDT&E										
Air Force	49.8	-	62.9	-	77.8	-	-	-	77.8	0
Navy	3.6	-	2.6	-	2.9	-	-	-	2.9	0
Subtotal	53.4	-	65.5	-	80.7	-	-	-	80.7	0
Procurement										0
Air Force	272.7	170	355.4	246	309.6	218	-	-	309.6	218
Navy	138.1	71	155.6	101	188.5	161	-	-	188.5	161
Subtotal	410.8	241	511.0	347	498.1	379	-	-	498.1	379
Spares	2.8	-	0.6	-	0.7	-	-	-	0.7	-
Total	467.0	241	577.1	347	579.5	379	-	-	579.5	379

* FY 2010 & FY 2011 include Base and OCO funding
** Reflects the FY 2011 President's Budget Request

Numbers may not add due to rounding

FY 2012 Program Acquisition Costs by Weapon System

Air Intercept Missile – 9X

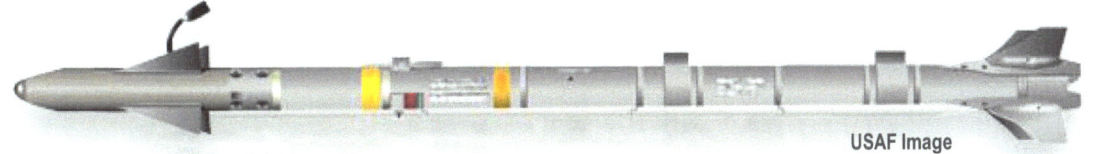

The Air Intercept Missile-9X (AIM-9X), also known as SIDEWINDER, is a short range air-to-air missile that provides a launch-and-leave air combat missile, which uses passive infrared energy for acquisition and tracking of enemy aircraft. The AIM-9X retains several components from the previous Sidewinder generation, the AIM-9M (primarily the motor and warhead), but incorporates a new airframe with much smaller fins and canards, and relies in a jet-vane steering system for significantly enhanced agility. The new guidance unit incorporates an advance Imaging Infrared (IIR) seeker. The AIM-9X is a joint Navy/Air Force program led by the Navy.

Mission: The mission of the AIM-9X is to destroy low and high altitude, high-speed enemy targets in an electronic countermeasures environment.

FY 2012 Program: Continues full rate production as well as product improvements, such as data link capabilities, and battery and safety improvements.

Prime Contractor: Raytheon Company, Tucson, AZ

Air Intercept Missile – 9X

	FY 2010*		FY 2011**		FY 2012 Base Budget		FY 2012 OCO Budget		FY 2012 Total Request	
	$M	Qty	$M	Qty	$M	Qty	$M	Qty	$M	Qty
RDT&E										
Air Force	5.9	-	6.0	-	8.0	-	-	-	8.0	-
Navy	2.2	-	1.0	-	8.8	-	-	-	8.8	-
Subtotal	8.1	-	7.0	-	16.8	-	-	-	16.8	-
Procurement										-
Air Force	78.5	219	64.5	178	88.8	240	-	-	88.8	240
Navy	53.7	161	55.2	155	47.1	132	-	-	47.1	132
Subtotal	132.2	380	119.7	333	135.9	372	-	-	135.9	372
Spares	1.1	-	0.9	-	0.8	-	-	-	0.8	-
Total	141.4	380	127.6	333	153.5	372	-	-	153.5	372

* FY 2010 & FY 2011 include Base and OCO funding
** Reflects the FY 2011 President's Budget Request

Numbers may not add due to rounding

FY 2012 Program Acquisition Costs by Weapon System

Chemical Demilitarization

DOD - JOINT

The Chemical Demilitarization Program is composed of two Major Defense Acquisition Programs, which are the U.S. Army Chemical Weapons Agency (CMA) and the Assembled Chemical Weapons Alternatives (ACWA) Program, with the goal of destroying a variety of chemical agents and weapons, including the destruction of former chemical weapon production facilities. This program is designed to eliminate the existing chemical weapons stockpile in compliance with the Chemical Weapons Convention (CWC) signed in 1997 – while ensuring the safety and security of the workers, the public, and the environment. Under the CWC, the United States has an obligation to destroy the chemical weapons stockpile by April 29, 2012.

US Army Photo

Mission: There are five mission areas within the Chemical Demilitarization Program:
1. Destroy chemical agents and weapons stockpile using incineration technology;
2. Destroy bulk container chemical agents stockpiles using neutralization technology;
3. Destroy chemical agents and weapons stockpiles using neutralization technologies;
4. Destroy Chemical Warfare Material (CWM) apart from the stockpile including: disposal of binary chemical weapons, former production facilities, and recovered chemical weapons; and
5. Chemical stockpile emergency preparedness.

FY 2012 Program: Continues safe and secure destruction operations at the three CMA operating sites (Toole, UT, Anniston, AL, and Umatilla, OR) with a goal of 90 percent destruction of the U.S. chemical weapons by 2012; continues closure activities at one site (Pine Bluff, AR); and begins closure activities at the three remaining CMA sites (Toole, UT, Anniston, AL, and Umatilla, OR). Funds ongoing construction efforts at the ACWA Program sites (Pueblo, CO and Blue Grass, KY) to accelerate completing destruction of the remaining 10 percent of the U.S. chemical stockpile as close to 2017 as possible, in accordance with the National Defense Authorization Act (NDAA) for FY 2008.

Prime Contractors: URS Corporation, Arlington, VA; Bechtel National Incorporated, Pueblo, CO; Bechtel Parsons, Richmond, KY

Chemical Demilitarization

	FY 2010*		FY 2011**		FY 2012 Base Budget		FY 2012 OCO Budget		FY 2012 Total Request	
	$M	Qty	$M	Qty	$M	Qty	$M	Qty	$M	Qty
Chemical Agents and Munitions Destruction	1,560.8		1,467.3		1,554.4				1,554.4	-
MILCON	151.5		124.7		75.3				75.3	-
Total	1,712.3	-	1,592.0	-	1,629.7	-	-	-	1,629.7	-

* FY 2010 & FY 2011 include Base and OCO funding
** Reflects the FY 2011 President's Budget Request

Numbers may not add due to rounding

FY 2012 Program Acquisition Costs by Weapon System

Joint Air-to-Ground Missile

The Joint Air-To-Ground Missile (JAGM) is a joint Army and Navy program led by the Army to provide a conventional, precision-guided, air-to-ground weapon that can be delivered from both fixed and rotary wing aircraft. The JAGM is intended to replace the aging inventory of Hellfire and Maverick missiles. The concept of JAGM is to employ a multi-mode seeker to attack fixed and moving targets alike.

Mission: The mission of JAGM is to provide close air support with the ability to attack fixed and moving targets.

FY 2012 Program: Continues system development.

Prime Contractor: Currently in Source Selection

Joint Air-to-Ground Missile

	FY 2010*		FY 2011**		FY 2012 Base Budget		FY 2012 OCO Budget		FY 2012 Total Request	
	$M	Qty	$M	Qty	$M	Qty	$M	Qty	$M	Qty
RDT&E										
Army	118.5	-	130.3	-	127.1	-	-	-	127.1	-
Navy	61.8	-	100.8	-	118.4	-	-	-	118.4	-
Subtotal	180.3	-	231.1	-	245.5	-	-	-	245.5	-
Procurement	-	-	-	-	-	-	-	-	-	-
Spares	-	-	-	-	-	-	-	-	-	-
Total	180.3	-	231.1	-	245.5	-	-	-	245.5	-

* FY 2010 & FY 2011 include Base and OCO funding
** Reflects the FY 2011 President's Budget Request

Numbers may not add due to rounding

FY 2012 Program Acquisition Costs by Weapon System

Joint Air to Surface Standoff Missile

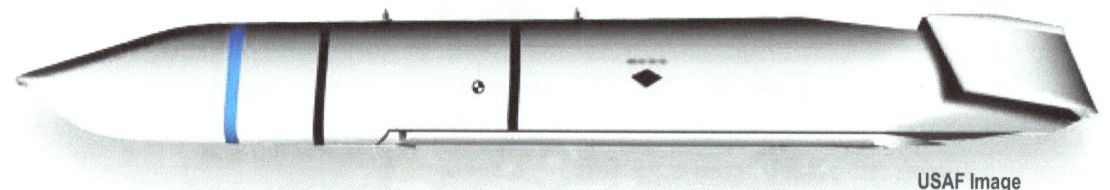
USAF Image

The Joint Air-to-Surface Standoff Missile (JASSM) is a joint Air Force and Navy program led by the Air Force to provide a conventional precision guided, long range standoff cruise missile that can be delivered from both fighters and bombers. The JASSM is procured in a baseline variant as well as an extended range (JASSM-ER) variant. Only the Air Force is currently buying this weapon.

Mission: The mission of the JASSM is to destroy targets from a long-range standoff position deliverable by fighter and bomber aircraft.

FY 2012 Program: Continues full rate production that was resumed in FY 2011. The JASSM program had a production pause in FY 2010 due to technical issues. Given successful testing since of both baseline and ER variants, JASSM will resume production in FY 2011, and continue in FY 2012.

Prime Contractor: Lockheed Martin Corporation, Troy, AL

Joint Air to Surface Standoff Missile

	FY 2010*		FY 2011**		FY 2012					
					Base Budget		OCO Budget		Total Request	
	$M	Qty	$M	Qty	$M	Qty	$M	Qty	$M	Qty
RDT&E	28.5	-	20.0	-	5.8	-	-	-	5.8	-
Procurement	52.5	-	215.8	171	236.2	142	-	-	236.2	142
Spares	-	-	-	-	-	-	-	-	-	-
Total	81.0	-	235.8	171	242.0	142	-	-	242.0	142

* FY 2010 & FY 2011 include Base and OCO funding
** Reflects the FY 2011 President's Budget Request

Numbers may not add due to rounding

FY 2012 Program Acquisition Costs by Weapon System

Joint Direct Attack Munition

The Joint Direct Attack Munition (JDAM) is a joint Air Force and Navy program led by the Air Force. The JDAM improves the existing inventory of general purpose gravity bombs by integrating a Global Positioning System (GPS)/inertial navigation guidance capability that improves accuracy and adverse weather capability.

Mission: This program enhances DoD conventional strike system capabilities by providing the ability to precisely attack time-critical, high value fixed or maritime targets under adverse environmental conditions and from all altitudes.

FY 2012 Program: Continues production of the system at low rate, given acceptable inventory levels of JDAM.

Prime Contractor: The Boeing Company, St. Charles, MO

Joint Direct Attack Munition

	FY 2010*		FY 2011**		FY 2012					
					Base Budget		OCO Budget		Total Request	
	$M	Qty	$M	Qty	$M	Qty	$M	Qty	$M	Qty
RDT&E	50.0	-	-	-	-	-	-	-	-	-
Procurement										
Air Force	190.4	7,517	252.6	9,331	76.6	3,250	34.1	1,338	110.7	4,588
Navy	1.9	-	-	-	-	-	-	-	-	-
Subtotal	192.3	7,517	252.6	9,331	76.6	3,250	34.1	1,338	110.7	4,588
Spares	-	-	-	-	-	-	-	-	-	-
Total	242.3	7,517	252.6	9,331	76.6	3,250	34.1	1,338	110.7	4,588

FY 2010 & FY 2011 include Base and OCO funding
** Reflects the FY 2011 President's Budget Request

Numbers may not add due to rounding

Joint Standoff Weapon

The Joint Standoff Weapon (JSOW - AGM-154) program is a joint Navy and Air Force program led by the Navy. The JSOW provides day, night, and adverse weather environment munition capability, and consists of three variants. The JSOW baseline (BLU-97) provides a day/night all-weather environment submunition for soft and area targets. The JSOW anti-armor variant (BLU-108) contains precision-guided anti-armor submunition warheads. The JSOW Unitary incorporates the dual-stage Broach penetrating warhead with terminal accuracy via Automatic Target Acquisition Seeker Technology.

Mission: The JSOW is a primary standoff precision guided munition. The day/night, adverse weather capability provides continuous munitions operations from a survivable standoff range. The Air Force stopped production of JSOW in FY 2005, favoring other weapons to meet the requirement.

FY 2012 Program: Continues production and product improvements of JSOW Unitary for the Navy only.

Prime Contractor: Raytheon Company, Tucson, AZ

Joint Standoff Weapon

	FY 2010*		FY 2011**		FY 2012 Base Budget		FY 2012 OCO Budget		FY 2012 Total Request	
	$M	Qty	$M	Qty	$M	Qty	$M	Qty	$M	Qty
RDT&E	9.7	-	12.6	-	7.5	-	-	-	7.5	-
Procurement	142.0	313	131.1	333	137.7	266	-	-	137.7	266
Spares	0.2	-	0.2	-	0.2	-	-	-	0.2	-
Total	151.9	313	143.9	333	145.4	266	-	-	145.4	266

* FY 2010 & FY 2011 include Base and OCO funding
** Reflects the FY 2011 President's Budget Request

Numbers may not add due to rounding

FY 2012 Program Acquisition Costs by Weapon System

Small Diameter Bomb — DOD - JOINT

The Small Diameter Bomb (SDB) is a joint Air Force and Navy program led by the Air Force to provide a conventional small sized, precision guided, standoff air-to-ground weapon that can be delivered from both fighters and bombers. The SDB-I was a fixed-target attack weapon, whereas the SDB-II incorporates a seeker and data link for use against fixed targets.

USAF Image

Mission: The mission of the SDB is to destroy targets from a medium-range standoff position deliverable by both fighters and bombers, with higher loadout and less collateral damage compared to other weapons.

FY 2012 Program: Procures only a Focused Lethality Munition variant of SDB, similar to SDB-I, but a variant that increases the near field blast while decreasing collateral damage further. Beginning the following year (FY 2013), the planned procurement will be SDB-II.

Prime Contractor: Raytheon Missile Systems, Tucson, AZ (SDB-I)
Boeing St. Charles, MO (SDB-II & FLM)

	FY 2010*		FY 2011**		FY 2012 Base Budget		FY 2012 OCO Budget		FY 2012 Total Request	
	$M	Qty	$M	Qty	$M	Qty	$M	Qty	$M	Qty
RDT&E										
Air Force	150.1	-	153.5	-	132.9	-	-	-	132.9	-
Navy	17.5	-	44.1	-	47.6	-	-	-	47.6	-
Subtotal	167.6	-	197.6	-	180.5	-	-	-	180.5	-
Procurement										-	-
Air Force	141.7	2,694	134.9	2,985	7.5	-	12.3	100	19.8	100
Spares	-	-	-	-	-	-	-	-	-	-
Total	309.3	2,694	332.5	2,985	188.0	-	12.3	100	200.3	100

* FY 2010 & FY 2011 include Base and OCO funding
** Reflects the FY 2011 President's Budget Request

Numbers may not add due to rounding

FY 2012 Program Acquisition Costs by Weapon System

Javelin Advanced Anti-Tank Weapon

USMC Photo

The Javelin Advanced Anti-tank Weapon System-Medium is a man-portable fire-and-forget weapon system used against tanks with conventional and reactive armor. Special features of Javelin are the choice of top attack or direct fire mode, integrated day/night sight, soft launch permitting fire from enclosures, and imaging infrared seeker.

Mission: To defeat armored targets with a man-portable weapon.

FY 2012 Program: Continues full rate production of missiles, Command Launch Units (CLU), and training devices.

Prime Contractor: Raytheon/Lockheed Martin Javelin Joint Venture, Tucson, AZ and Orlando, FL

Javelin Advanced Anti-Tank Weapon

	FY 2010*		FY 2011**		FY 2012 Base Budget		FY 2012 OCO Budget		FY 2012 Total Request	
	$M	Qty	$M	Qty	$M	Qty	$M	Qty	$M	Qty
RDT&E	-	-	10.0	-	17.3	-	-	-	17.3	-
Procurement	258.6	1,334	163.9	715	160.8	710	-	-	160.8	710
Spares	-	-	-	-	-	-	-	-	-	-
Total	258.6	1,334	173.9	715	178.1	710	-	-	178.1	710

* FY 2010 & FY 2011 include Base and OCO funding
** Reflects the FY 2011 President's Budget Request

Numbers may not add due to rounding

MUNITIONS AND MISSILES

FY 2012 Program Acquisition Costs by Weapon System

Guided Multiple Launch Rocket System

US Army Photo

The Guided Multiple Launch Rocket System (GMLRS) consists of a C-130 transportable, wheeled, indirect fire, rocket/missile system capable of firing all rockets and missiles in the current and future Multiple Launch Rocket System (MLRS) family of munitions.

Mission: The mission of GMLRS is to neutralize or suppress enemy field artillery and air defense systems and supplement cannon artillery fires.

FY 2012 Program: Continues full rate production as well as product improvements such as insensitive munition and alternative warhead development for the Army.

Prime Contractor: Lockheed Martin Corporation, Dallas, TX

Guided Multiple Launch Rocket System

	FY 2010*		FY 2011**		FY 2012					
					Base Budget		OCO Budget		Total Request	
	$M	Qty	$M	Qty	$M	Qty	$M	Qty	$M	Qty
RDT&E	26.6	-	51.6	-	66.6	-	-	-	66.6	-
Procurement	353.3	3,228	291.0	2,592	314.2	2,784	19.0	210	333.2	2,994
Spares	-	-	-	-	-	-	-	-	-	-
Total	379.9	3,228	342.6	2,592	380.8	2,784	19.0	210	399.8	2,994

* FY 2010 & FY 2011 include Base and OCO funding
** Reflects the FY 2011 President's Budget Request

Numbers may not add due to rounding

FY 2012 Program Acquisition Costs by Weapon System

Evolved Seasparrow Missile

NSPO Photo

The Evolved Seasparrow Missile (ESSM) is an improved version of the NATO Seasparrow missile, designed for ship self-defense.

Mission: The mission of the ESSM is to provide to the Navy a missile with performance to defeat current and projected threats that possess low altitude, high velocity, and maneuver characteristics beyond the engagement capabilities of other ship self-defense systems.

FY 2012 Program: Continues full rate production.

Prime Contractor: Raytheon Company, Tucson, AZ

Evolved Seasparrow Missile

	FY 2010*		FY 2011**		FY 2012					
					Base Budget		OCO Budget		Total Request	
	$M	Qty	$M	Qty	$M	Qty	$M	Qty	$M	Qty
Procurement	51.2	43	48.2	33	48.5	35	-	-	48.5	35
Spares	-	-	-	-	-	-	-	-	-	-
Total	51.2	43	48.2	33	48.5	35	-	-	48.5	35

* FY 2010 & FY 2011 include Base and OCO funding
** Reflects the FY 2011 President's Budget Request

Numbers may not add due to rounding

FY 2012 Program Acquisition Costs by Weapon System

Rolling Airframe Missile

The Rolling Airframe Missile (RAM) is a high firepower, Lightweight complementary self-defense system to engage anti-ship cruise missiles.

Mission: The mission of the RAM is to provide high firepower close-in defense of combatant and auxiliary ships by utilizing a dual mode, passive radio frequency/infrared missile in a compact 21 missile launcher.

FY 2012 Program: Continues production of missiles and alterations

Prime Contractor: Raytheon Company, Tucson, AZ

Rolling Airframe Missile

	FY 2010*		FY 2011**		FY 2012 Base Budget		FY 2012 OCO Budget		FY 2012 Total Request	
	$M	Qty	$M	Qty	$M	Qty	$M	Qty	$M	Qty
Procurement	69.7	90	75.0	90	66.2	61	-	-	66.2	61
Spares	-	-	-	-	-	-	-	-	-	-
Total	69.7	90	75.0	90	66.2	61	-	-	66.2	61

* FY 2010 & FY 2011 include Base and OCO funding
** Reflects the FY 2011 President's Budget Request

Numbers may not add due to rounding

MUNITIONS AND MISSILES

FY 2012 Program Acquisition Costs by Weapon System

Standard Family of Missiles

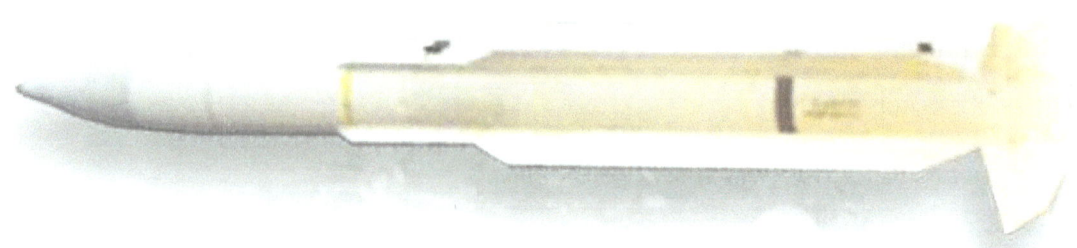
US Navy Photo

The STANDARD missile family consists of various air defense missiles including supersonic, medium, and extended range; surface-to-air; and surface-to-surface missiles. The Standard Missile-6 is a surface Navy Anti-Air Warfare (AAW) missile that provides area and ship self defense. The missile is intended to project power and contribute to raid annihilation by destroying manned fixed and rotary wing aircraft, Unmanned Aerial Vehicles (UAV), Land Attack Cruise Missiles (LACM), and Anti-Ship Cruise Missiles (ASCM) in flight. It was designed to fulfill the need for a vertically launched, extended range missile compatible with the Aegis Weapon System (AWS) to be used against extended range threats at-sea, near land, and overland. SM-6 combines the tested legacy of STANDARD Missile -2 (SM-2) propulsion and ordnance with an active Radio Frequency (RF) seeker modified from the AIM-120 Advanced Medium Range Air-to-Air Missile (AMRAAM), allowing for over-the-horizon engagements, enhanced capability at extended ranges and increased firepower.

Mission: The mission of the STANDARD missile family is to provide all-weather, anti-aircraft and surface-to-surface armament for cruisers, destroyers, and guided missile frigates. The most recent variant of Standard Missile is SM-6, which incorporated an AMRAAM seeker for increased performance, including overland capability.

FY 2012 Program: Continues production of the SM-6 variant.

Prime Contractor: Raytheon Company, Tucson, AZ

Standard Family of Missiles

	FY 2010*		FY 2011**		FY 2012					
					Base Budget		OCO Budget		Total Request	
	$M	Qty	$M	Qty	$M	Qty	$M	Qty	$M	Qty
RDT&E	150.1	-	96.2	-	46.7	-	-	-	46.7	-
Procurement	188.5	45	295.9	67	420.3	89	-	-	420.3	89
Spares	-	-	-	-	-	-	-	-	-	-
Total	338.6	45	392.1	67	467.0	89	-	-	467.0	89

* FY 2010 & FY 2011 include Base and OCO funding
** Reflects the FY 2011 President's Budget Request

Numbers may not add due to rounding

FY 2012 Program Acquisition Costs by Weapon System

Tactical Tomahawk Cruise Missile

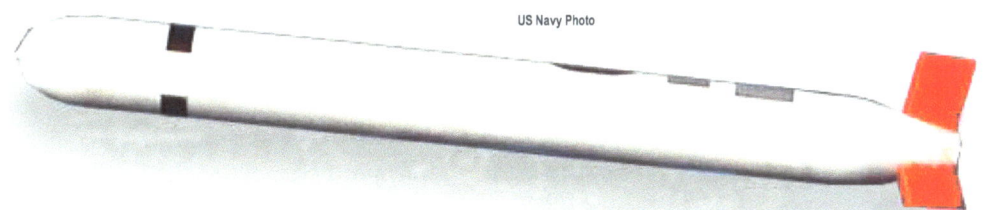

US Navy Photo

The Tactical Tomahawk is a Navy cruise missile weapon system with a long-range conventional warhead system, which is sized to fit torpedo tubes and capable of being deployed from a variety of surface ship and submarine platforms. The Tactical Tomahawk, also referred to as Block IV, incorporates an active electronically scanned array, millimeter-wave seeker, which provides target acquisition and homing; and a passive electronic surveillance system is for long-range acquisition and identification. The missile carries a 1,000-lb. warhead, and is normally launched from a SSNs equipped with the vertical launch systems. The Block IV missiles also provides new capability enhancements, to include increased flexibility utilizing two-way satellite communications to reprogram the missile in-flight, and increased responsiveness with faster launch timelines.

Mission: The mission of the TOMAHAWK is to provide a long-range cruise missile launched from a variety of platforms against land and sea targets.

FY 2012 Program: Continues production at a minimum sustaining rate.

Prime Contractor: Raytheon Company, Tucson, AZ

Tactical Tomahawk Cruise Missile

	FY 2010*		FY 2011**		FY 2012					
					Base Budget		OCO Budget		Total Request	
	$M	Qty	$M	Qty	$M	Qty	$M	Qty	$M	Qty
RDT&E	16.7	-	10.6	-	8.8	-	-	-	8.8	-
Procurement	276.5	196	300.2	196	303.3	196	-	-	303.3	196
Spares	-	-	-	-	-	-	-	-	-	-
Total	293.2	196	310.8	196	312.1	196	-	-	312.1	196

* FY 2010 & FY 2011 include Base and OCO funding
** Reflects the FY 2011 President's Budget Request

Numbers may not add due to rounding

FY 2012 Program Acquisition Costs by Weapon System

Trident II Ballistic Missile Modifications

The Trident II (D5) is a submarine launched ballistic missile with greater range, payload capability, and accuracy than the Trident I (C4) missile.

Mission: The mission of the Trident II (D5) ballistic missile is to deter nuclear war by means of assured retaliation in response to a major attack on the United States or its Allies, and to enhance nuclear stability by providing no incentive for enemy first strike. The Trident II (D5) missile has the ability to precisely attack time-critical, high value, fixed targets. The importance of this program as a key component to the sea-based leg of the nuclear triad was confirmed by the President and Congress in the New START Treaty ratification.

FY 2012 Program: Funds the D5 Missile Life Extension Program replacing missile motors and other critical components, and production support (including flight test instrumentation and additional re-entry system hardware).

Prime Contractor: Lockheed Martin Corporation, Sunnyvale, CA

US Navy Photo

Trident II Ballistic Missile

	FY 2010*		FY 2011**		FY 2012 Base Budget		FY 2012 OCO Budget		FY 2012 Total Request	
	$M	Qty	$M	Qty	$M	Qty	$M	Qty	$M	Qty
RDT&E	68.0		81.2		88.9				88.9	
Procurement	1,046.7	24	1,106.9	24	1,309.1	24			1,309.1	24
Spares									-	-
Total	1,114.7	24	1,188.1	24	1,398.0	24	-	-	1,398.0	24

* FY 2010 & FY 2011 include Base and OCO funding
** Reflects the FY 2011 President's Budget Request

Numbers may not add due to rounding

Shipbuilding and Maritime Systems

A central principle to the U.S. Maritime Strategy is forward presence. Forward presence promotes conflict deterrence by ensuring forces are in a position to expeditiously respond to conflict. Therefore, sea services must buy, build, and maintain maritime systems in accordance with mission need.

The Shipbuilding Portfolio details programs that ensure the accomplishment of the overall maritime mission. The Shipbuilding Portfolio consists of bought, built, and maintained systems, subsystems, and components. The 313-ship fleet will allow the U.S. to maintain maritime superiority well into the 21st century. The mobilization of the 313-ship fleet will ensure mission accomplishment. The following highlights the FY 2012 Shipbuilding Portfolio budget request:

FY 2012 Shipbuilding and Maritime Systems – Base and OCO: $24.2 Billion

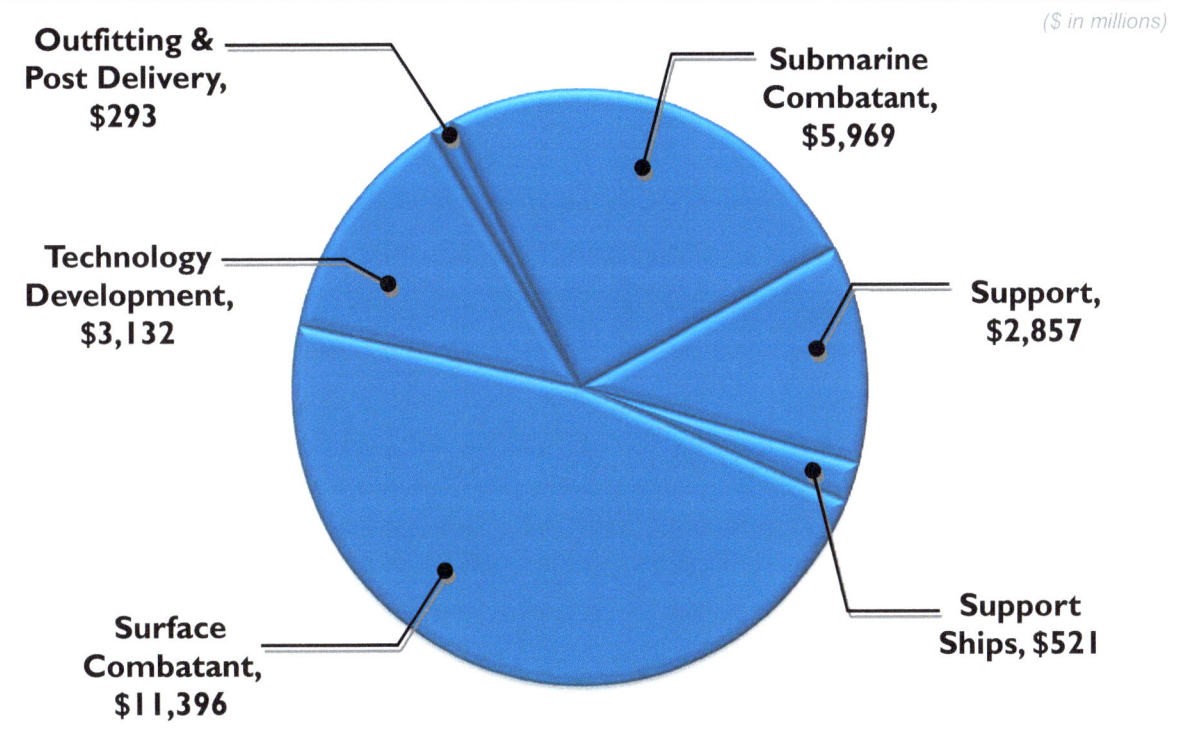

($ in millions)

- Outfitting & Post Delivery, $293
- Submarine Combatant, $5,969
- Technology Development, $3,132
- Support, $2,857
- Surface Combatant, $11,396
- Support Ships, $521

Source: FY 2012 PRCP – Investment Categorization
Support ship subcategory includes MLP ship in the National Defense Sealift Fund (NDSF)
Numbers may not add due to rounding

FY 2012 Program Acquisition Costs by Weapon System

Joint High Speed Vessel — DOD - JOINT

The Joint High Speed Vessel (JHSV) is a cooperative Army and Navy effort for a high speed shallow draft vessel designed for rapid intra-theater transport.

Mission: The JHSVs provide combatant commanders with high-speed, intra-theater sealift mobility, inherent cargo handling capacity, and the agility to achieve positional advantage over operational distances. Delivery of the first JHSV is scheduled for the first quarter of FY 2013.

FY 2012 Program: Funds two JHSV ships, one each for Army and Navy, which are predominantly commercially designed vessels that will cost approximately $204.5 million each.

Prime Contractor: Austal USA, Mobile, AL

US Navy Image

Joint High Speed Vessel

	FY 2010*		FY 2011**		FY 2012 Base Budget		FY 2012 OCO Budget		FY 2012 Total Request	
	$M	Qty	$M	Qty	$M	Qty	$M	Qty	$M	Qty
RDT&E										
Navy	8.2	-	3.6	-	4.1	-	-	-	4.1	-
Army	3.0	-	3.2	-	3.0	-	-	-	3.0	-
Subtotal	11.2	-	6.8	-	7.1	-	-	-	7.1	-
Procurement										-	-
Navy	177.4	1	180.7	1	185.1	1	-	-	185.1	1
Army	202.5	1	202.8	1	223.8	1	-	-	223.8	1
Subtotal	379.9	2	202.8	2	408.9	2	-	-	408.9	2
Spares	-	-	-	-	-	-	-	-	-	-
Total	391.1	2	209.6	2	416.0	2	-	-	416.0	2

* FY 2010 & FY 2011 reflect Base, No OCO funding
** Reflects the FY 2011 President's Budget Request

Numbers may not add due to rounding

SHIPBUILDING AND MARITIME SYSTEMS

FY 2012 Program Acquisition Costs by Weapon System

DDG 51 AEGIS Destroyer

The DDG 51 AEGIS Destroyer Class ships operate defensively and offensively as units of Carrier Strike Groups and Surface Action Groups, in support of Underway Replenishment Groups and the Marine Amphibious Task Forces in multi-threat environments, which include air, surface, and subsurface threats. The DDG 51 ship is armed with a vertical launching system, which accommodates 96 missiles and a 5 inch gun that provides Naval Surface Fire Support to forces ashore and anti-ship gunnery capability.

US Navy Photo

Mission: The DDG 51 AEGIS Destroyer ship provides air and maritime dominance and land attack capability with its Aegis Anti-Submarine and Tomahawk Weapon Systems. The DDG 51 Flight III will meet ballistic missile defense and open ocean anti-submarine warfare (ASW) requirements.

FY 2012 Program: Funds one DDG 51 AEGIS Destroyer.

Prime Contractors: General Dynamics Corporation, Bath, ME
Northrop Grumman Corporation, Pascagoula, MS

DDG 51 AEGIS Destroyer

	FY 2010*		FY 2011**		FY 2012					
					Base Budget		OCO Budget		Total Request	
	$M	Qty	$M	Qty	$M	Qty	$M	Qty	$M	Qty
RDT&E	-	-	-	-	-	-	-	-	-	-
Procurement	2,483.6	1	2,970.2	2	2,081.4	1			2,081.4	1
Spares									-	-
Total	2,483.6	1	2,970.2	2	2,081.4	1	-	-	2,081.4	1

* FY 2010 & FY 2011 reflect Base, No OCO funding
** Reflects the FY 2011 President's Budget Request

Numbers may not add due to rounding

SHIPBUILDING AND MARITIME SYSTEMS

Littoral Combat Ship

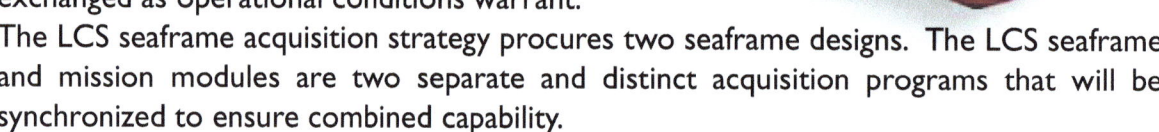

Image courtesy General Dynamics

Image Courtesy Lockheed Martin

The Littoral Combat Ship (LCS) will be a fast, agile, and stealthy surface combatant capable of anti-access missions against asymmetric threats in the littorals. It will be the first Navy ship to separate capability from hull form. For example, LCS will be capable of employing interchangeable mission modules for Mine Warfare, Anti-Submarine Warfare, and Anti-Surface Warfare to counter anti-access threats close to shore such as mines, quiet diesel submarines, and swarming small boats. The LCS mission modules will be exchanged as operational conditions warrant. The LCS seaframe acquisition strategy procures two seaframe designs. The LCS seaframe and mission modules are two separate and distinct acquisition programs that will be synchronized to ensure combined capability.

Mission: The LCS defeats asymmetric threats, and assures naval and joint forces access into contested littoral regions by prosecuting small boats, mines countermeasures, and littoral anti-submarine warfare.

FY 2012 Program: Funds four LCS seaframes at $1.9 billion and two mission modules (Mine Countermeasures and Surface Warfare) at $79.6 million.

Prime Contractors: Lockheed Martin, Marinette, WI and Austal USA, Mobile, AL

Littoral Combat Ship

	FY 2010*		FY 2011**		FY 2012					
					Base Budget		OCO Budget		Total Request	
	$M	Qty	$M	Qty	$M	Qty	$M	Qty	$M	Qty
RDT&E	422.0	-	226.3	-	286.8	-	-	-	286.8	-
Procurement***	1,157.1	2	1,592.3	2	1,881.7	4	-	-	1,881.7	4
Spares	-	-	-	-	-	-	-	-	-	-
Total	1,579.1	2	1,818.6	2	2,168.5	4	-	-	2,168.5	4

* FY 2010 & FY 2011 reflect Base, No OCO funding
** Reflects the FY 2011 President's Budget Request
***Includes other procurement for mission modules

Numbers may not add due to rounding

FY 2012 Program Acquisition Costs by Weapon System

LPD 17 Amphibious Transport Dock Ship

US Navy Photo

The San Antonio Class Amphibious Transport Dock ships (LPD 17) are functional replacements for 41 ships of 4 classes of amphibious ships. The LPD 17 design includes systems configurations that reduce operating and support costs, and other operational performance improvements. System engineering and integration efforts have developed further reductions in life-cycle costs and integrated performance upgrades in a rapid, affordable manner.

Mission: The LPD 17 San Antonio Class Amphibious Transport Dock ships embark, transport, and land Marines in amphibious assault by helicopters, landing crafts, and amphibious vehicles.

FY 2012 Program: Funds the final 11th ship and line shutdown cost.

Prime Contractors: Northrop Grumman, Pascagoula, MS and New Orleans, LA

LPD 17 Amphibious Transport Dock Ship

	FY 2010*		FY 2011**		FY 2012 Base Budget		OCO Budget		Total Request	
	$M	Qty	$M	Qty	$M	Qty	$M	Qty	$M	Qty
RDT&E	5.1	-	1.4	-	0.9	-	-	-	0.9	-
Procurement	1,152.7	-	-	-	1,847.4	1	-	-	1,847.4	1
Spares	-	-	-	-	-	-	-	-	-	-
Total	1,157.8	-	1.4	-	1,848.3	1	-	-	1,848.3	1

* FY 2010 & FY 2011 reflect Base, No OCO funding
** Reflects the FY 2011 President's Budget Request

Numbers may not add due to rounding

FY 2012 Program Acquisition Costs by Weapon System

SSN 774 Virginia Class Submarine

US Navy Photo

The Virginia Class Submarine is an attack submarine that provides the Navy with the capabilities to maintain undersea supremacy in the 21st century. The Virginia Class Submarine is nuclear-powered and is intended to replace the fleet of 688 class submarines. It is characterized by state-of-the-art stealth and enhanced features for Special Operations Forces. Virginia Class Submarines are able to attack targets ashore with Tomahawk cruise missiles and to conduct covert long-term surveillance of land areas, littoral waters, and other sea-based forces.

Mission: The Virginia Class Submarines seek and destroy enemy ships across a wide spectrum of scenarios, working independently and in consort with a battle group and other ships, providing joint commanders with early, accurate knowledge of the battlefield.

FY 2012 Program: Funds two ships at $3.2 billion as part of an existing multiyear procurement contract and advance procurement of $1.5 billion for two ships in FY 2013 and two ships in FY 2014.

Prime Contractors: General Dynamics Corporation, Groton, CT
Northrop Grumman Corporation, Newport News, VA

SSN 774 Virginia Class Submarine

	FY 2010*		FY 2011**		FY 2012					
					Base Budget		OCO Budget		Total Request	
	$M	Qty	$M	Qty	$M	Qty	$M	Qty	$M	Qty
RDT&E	177.0	-	155.5	-	97.2	-	-	-	97.2	-
Procurement***	4,057.4	1	5,264.7	2	4,857.7	2	-	-	4,857.7	2
Spares	-	-	-	-	-	-	-	-	-	-
Total	4,234.4	1	5,420.2	2	4,954.9	2	-	-	4,954.9	2

* FY 2010 & FY 2011 reflect Base, No OCO funding
** Reflects the FY 2011 President's Budget Request
*** Includes other procurement for support equipment

Numbers may not add due to rounding

FY 2012 Program Acquisition Costs by Weapon System

LHA Replacement

The Landing Helicopter Assault Replacement (LHA-R) amphibious assault ship is the largest of all amphibious warfare ships. The LHA-R enhances the aviation capability of the decommissioning TARAWA Class LHA that reaches the end of its extended service life by 2015. The LHA-R resembles a small aircraft carrier and will be compatible with the future Marine Aviation Combat Element.

US Navy Image

Mission: The LHA-R provides short take-off vertical landing and vertical take-off landing capability for Marine aviation in the Expeditionary Strike Group (ESG).

FY 2012 Program: Funds second increment of LHA-7.

Prime Contractor: Northrop Grumman Ship Systems, Pascagoula, MS

LHA Replacement

	FY 2010*		FY 2011**		FY 2012 Base Budget		FY 2012 OCO Budget		FY 2012 Total Request	
	$M	Qty	$M	Qty	$M	Qty	$M	Qty	$M	Qty
RDT&E	-	-	-	-	-	-	-	-	-	-
Procurement	169.5	-	949.9	1	2,018.7	-	-	-	2,018.7	-
Spares									-	-
Total	169.5	-	949.9	1	2,018.7	-	-	-	2,018.7	-

* FY 2010 & FY 2011 reflects Base, No OCO funding
** Reflects the FY 2011 President's Budget Request

Numbers may not add due to rounding

FY 2012 Program Acquisition Costs by Weapon System

Mobile Landing Platform

US Navy Image

The Mobile Landing Platform (MLP) interfaces with other ships at sea and surface connectors to transfer vehicles, personnel, and equipment for deployment ashore from the sea base. It is part of the Maritime Prepositioning Force (MPF) and serves as the principle interface of the organic surface connectors for the MPF squadron.

Mission: The MLP will provide enhanced at sea and surface connector vehicle, personnel, and equipment transfer capability for humanitarian and counter insurgency missions. It will also provide support to the Marine Expeditionary Brigade.

FY 2012 Program: Funds one MLP.

Prime Contractor: General Dynamics National Steel and Shipbuilding Company, San Diego, CA

MLP Replacement

| | FY 2010* | | FY 2011** | | FY 2012 | | | | | |
| | | | | | Base Budget | | OCO Budget | | Total Request | |
	$M	Qty	$M	Qty	$M	Qty	$M	Qty	$M	Qty
RDT&E	-	-	-	-	-	-	-	-	-	-
Procurement	120.0	-	380.0	1	425.9	1	-	-	425.9	1
Spares	-	-	-	-	-	-	-	-	-	-
Total	120.0	-	380.0	1	425.9	1	-	-	425.9	1

* FY 2010 & FY 2011 reflect Base, No OCO funding
** Reflects the FY 2011 President's Budget Request

Numbers may not add due to rounding

Space Based and Related Systems

Space assets support deployed United States forces by providing communications services, navigation capabilities, and information collected by remote sensors such as weather satellites and intelligence collection systems. Space forces contribute to the overall effectiveness of U.S. military forces by acting as a force multiplier that enhances combat power. The capability to control space contributes to achieving information superiority and battle space dominance.

Procurement of satellites and launch services are typically funded two years prior to launch. Generally speaking, the first two satellites of a new system are purchased with Research, Development, Test & Evaluation funding and the remainder of the satellites are purchased with procurement funding. The Air Force is implementing approaches to maximize efficient satellite acquisitions. These approaches include buying blocks of satellites, using fixed-price contracting to stabilize requirements, promoting a stable research and development investment for evolutionary growth, and modifying the annual funding approach for industrial base efficiency.

The FY 2012 overall space program request is higher this year at $10.2 billion than for FY 2011 (+3%), driven by an increased launch vehicle and satellite procurement tempo.

FY 2012 Space Based and Related Systems – Base and OCO: $10.2B

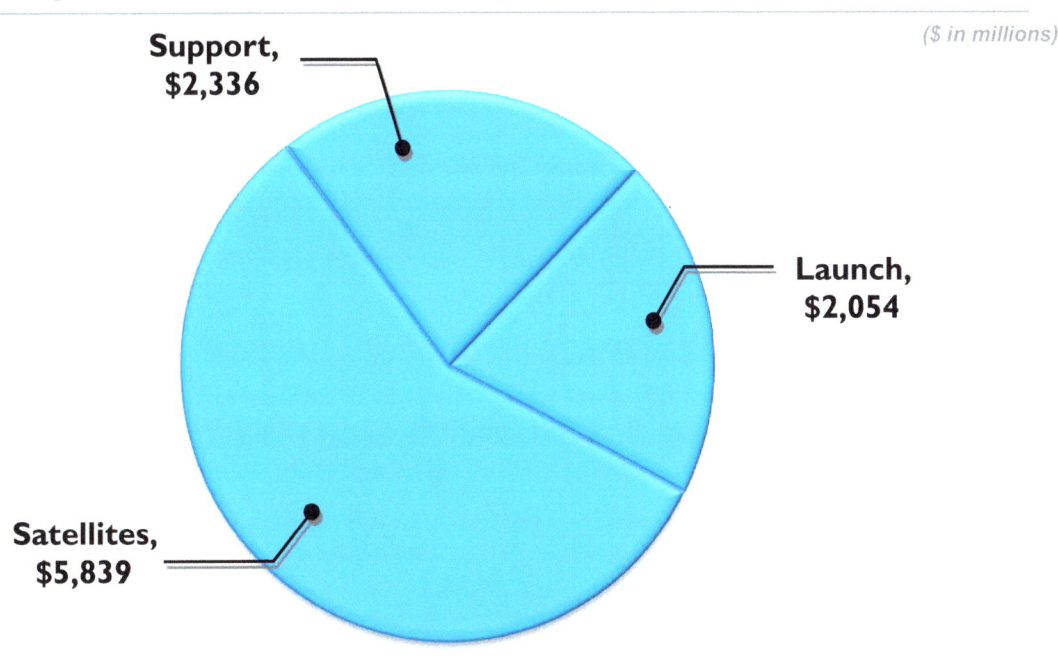

($ in millions)

- Support, $2,336
- Launch, $2,054
- Satellites, $5,839

Source: FY 2012 PRCP – Investment Categorization
Numbers may not add due to rounding

FY 2011 Program Acquisition Costs by Weapon System

Mobile User Objective System

The Mobile User Objective System (MUOS) is the next generation DoD advanced narrow band Ultra High Frequency (UHF) communications satellite constellation. It consists of four satellites in geosynchronous orbit with one on-orbit spare and a fiber optic terrestrial network connecting four ground stations. The MUOS satellite includes the new networked payload and a separate legacy payload. The MUOS will replace the existing UHF Follow-On (UFO) constellation and provide a much higher data rate capability for mobile users.

- There will be 16 beams per satellite with data rates of 64 kbps "on the move"
- The DoD Teleport will be the portal to the Defense Information System Network (DSN, SIPRNET and NIPRNET)
- The on–orbit capability for MUOS is planned for the first quarter of FY 2012

Mission: The MUOS will provide the mobile warfighter with point-to-point and netted communications services with a secure, "communications-on-the-move" capability on a 24 hours a day, 7 days a week basis.

FY 2012 Program: Funds procurement of the launch vehicle for satellite #4.

Prime Contractor: Lockheed Martin Corporation, Sunnyvale, CA

Image courtesy of Lockheed Martin

Mobile User Objective System

	FY 2010*		FY 2011**		FY 2012					
					Base Budget		OCO Budget		Total Request	
	$M	Qty	$M	Qty	$M	Qty	$M	Qty	$M	Qty
RDT&E	398.3		405.7		244.2				244.2	-
Procurement	509.9	1	505.7	1	238.2				238.2	-
Spares									-	-
Total	908.2	1	911.4	1	482.4	-	-	-	482.4	-

* FY 2010 & FY 2011 include Base and OCO funding
** Reflects the FY 2011 President's Budget Request

Numbers may not add due to rounding

FY 2011 Program Acquisition Costs by Weapon System

Advanced Extremely High Frequency

The Advanced Extremely High Frequency (AEHF) satellite will be a constellation of communications satellites in geosynchronous orbit that will replenish the existing EHF system MILSTAR satellite at a much higher capacity and data rate capability.

- 24-hour low, medium, and high data rate satellite connectivity from 65 N to 65 S latitude worldwide
- 8 full time spot beam antennas @ 75 bps to 8.192 Mbps data rate
- 24 time shared spot beam antennas @ 75 bps to 2.048 Mbps data rate
- 2 crosslink antennas per satellite (10 Mbps)
- Up to 160 cellular coverages (75 bps to 8.192 Mbps)
- X-band frequency data rate capable

The AEHF is a collaborative program that also includes resources for Canada, the United Kingdom, and the Netherlands.

Mission: The AEHF constellation will provide survivable, anti-jam, worldwide secure communications for strategic and tactical users.

FY 2012 Program: Funds SV-1 on-orbit tests and operations, SV-2 launch and on-orbit/operations support, Mission Control Segment (MCS) development. Propose the use of the Evolutionary Acquisition for Space Efficiency (EASE) approach to initiate the procurement of the SV 5-6 Block Buy, address obsolescence issues as well as implement a Capability and Affordability Improvement Program (CAIP) for future vehicles.

Prime Contractor: Lockheed Martin Corporation, Sunnyvale, CA

Advanced Extremely High Frequency

	FY 2010*		FY 2011**		FY 2012 Base Budget		FY 2012 OCO Budget		FY 2012 Total Request	
	$M	Qty	$M	Qty	$M	Qty	$M	Qty	$M	Qty
RDT&E	456.2	-	351.8	-	421.7	-	-	-	421.7	-
Procurement	1,836.7	1	246.6	-	552.8	2	-	-	552.8	2
Spares	-	-	-	-	-	-	-	-	-	-
Total	2,292.9	1	598.4	-	974.5	2	-	-	974.5	2

* FY 2010 & FY 2011 include Base and OCO funding
** Reflects the FY 2011 President's Budget Request

Numbers may not add due to rounding

FY 2011 Program Acquisition Costs by Weapon System

Evolved Expendable Launch Vehicle

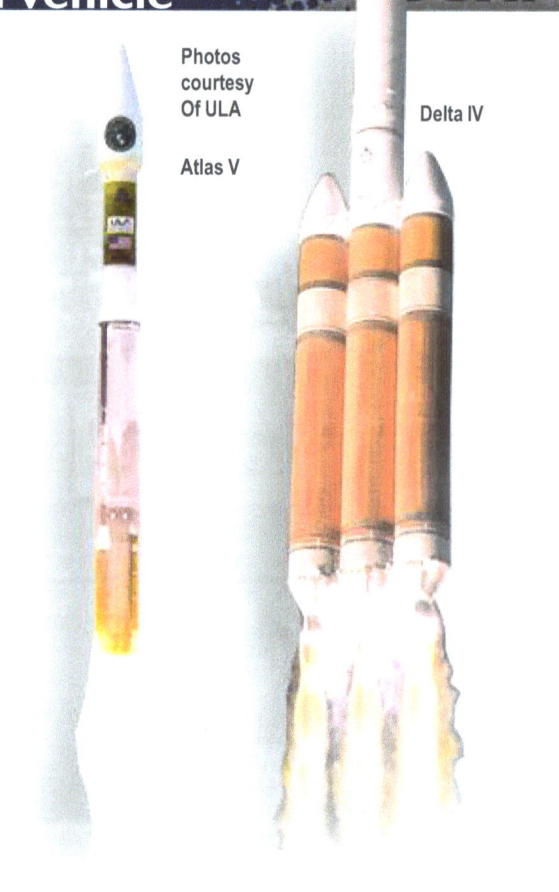

Photos courtesy Of ULA

Atlas V

Delta IV

The Evolved Expendable Launch Vehicle (EELV) replaced the heritage Delta, Atlas, and Titan launch vehicle families. The EELV provides the DoD, the National Reconnaissance Office (NRO), and other government and commercial purchasers launch services for medium to heavy lift class satellites. As of December 2006, the United Launch Alliance joint venture is the sole provider of EELV launch services.

Mission: The EELV program provides launch vehicles and services for medium and heavy class satellites.

FY 2012 Program: Funds the procurement of 4 launch vehicles and associated launch services and support activities. The figures below do not include EELVs for the Navy or NRO. Those launch vehicles are funded in the specific satellite program budgets.

Prime Contractor: United Launch Alliance, Decatur, AL

Evolved Expendable Launch Vehicle

	FY 2010*		FY 2011**		FY 2012					
					Base Budget		OCO Budget		Total Request	
	$M	Qty	$M	Qty	$M	Qty	$M	Qty	$M	Qty
RDT&E	43.9	-	30.2	-	20.0	-	-	-	20.0	-
Procurement	1,094.8	3	1,154.0	3	1,740.2	4	-	-	1,740.2	4
Spares	-	-	-	-	-	-	-	-	-	-
Total	1,138.7	3	1,184.2	3	1,760.2	4	-	-	1,760.2	4

* FY 2010 & FY 2011 include Base and OCO funding
** Reflects the FY 2011 President's Budget Request

Numbers may not add due to rounding

FY 2011 Program Acquisition Costs by Weapon System

Global Positioning System

The Global Positioning System (GPS) provides a global, three-dimensional positioning, navigation, and timing information system for aircraft, artillery, ships, tanks and other weapons delivery systems. The fully operational GPS constellation consists of at least 24 satellites on-orbit at all times. The GPS IIIA space vehicles will deliver significant enhancements, including a new L1C (civil) Galileo-compatible signal, enhanced M-code earth coverage power, and a growth path to full warfighter capabilities. Initial launch is planned for 2014.

Image Courtesy of Lockheed Martin

Mission: The GPS constellation provides worldwide positioning, navigation, and precise time to military and civilian users.

FY 2012 Program: Funds sustain the GPS constellation with the assembly and launch of replenishment satellites. Continues the development and production of the GPS IIIA system, the next generation GPS satellite, as well as development of the ground control system.

Prime Contractors: GPS IIIA: Lockheed Martin Corporation, King of Prussia, PA
GPS OCX Phase A: Raytheon Company, Aurora, CO

Global Positioning System

	FY 2010*		FY 2011**		FY 2012					
					Base Budget		OCO Budget		Total Request	
	$M	Qty	$M	Qty	$M	Qty	$M	Qty	$M	Qty
RDT&E	749.4	-	862.7	-	872.0	-	-	-	872.0	-
Procurement	131.0	-	194.8	-	590.0	2	-	-	590.0	2
Spares	-	-	-	-	-	-	-	-	-	-
Total	880.4	-	1,057.5	-	1,462.0	2	-	-	1,462.0	2

* FY 2010 & FY 2011 include Base and OCO funding
** Reflects the FY 2011 President's Budget Request

Numbers may not add due to rounding

7-5   SPACE BASED AND RELATED SYSTEMS

FY 2011 Program Acquisition Costs by Weapon System

Defense Weather Satellite System

Defense Weather Satellite System (DWSS) is the DoD component of the restructured NPOESS program. DWSS will satisfy DoD's environmental monitoring requirements in the early morning orbit by developing and launching two satellites, each with a Visible Infrared Imager Radiometer Suite (VIIRS), Space Environment Monitor (SEM-N), and Microwave Imager/Sounder (MIS) with an initial launch capability no earlier than 2018.

Mission: The DWSS will collect worldwide environmental data to support weather and oceanographic forecasting for military operational planning and protection of civilian life and property.

FY 2012 Program: Begins redesign of NPOESS spacecraft bus to a smaller and lighter version for DWSS. Continues development of VIIRS and MIS sensors, spacecraft and sensor subsystems and materials, algorithms, and DoD-specific elements of the common ground system. Starting in FY12 all procurement funds have been moved to Research Development Test & Evaluation due to the NPOESS restructure. NPOESS funds will transfer to the DWSS program element starting in the FY11 year of execution.

Prime Contractor: Northrop Grumman Corporation, Redondo Beach, CA

Defense Weather Satellite System

	FY 2010*		FY 2011**		FY 2012					
					Base Budget		OCO Budget		Total Request	
	$M	Qty	$M	Qty	$M	Qty	$M	Qty	$M	Qty
RDT&E	395.0	-	325.5	-	444.9	-	-	-	444.9	-
Procurement	3.9	-	26.3	-	-	-	-	-	-	-
Spares	-	-	-	-	-	-	-	-	-	-
Total	398.9	-	351.8	-	444.9	-	-	-	444.9	-

* FY 2010 & FY 2011 include Base and OCO funding
** Reflects the FY 2011 President's Budget Request

Numbers may not add due to rounding

FY 2011 Program Acquisition Costs by Weapon System

Space Based Infrared System

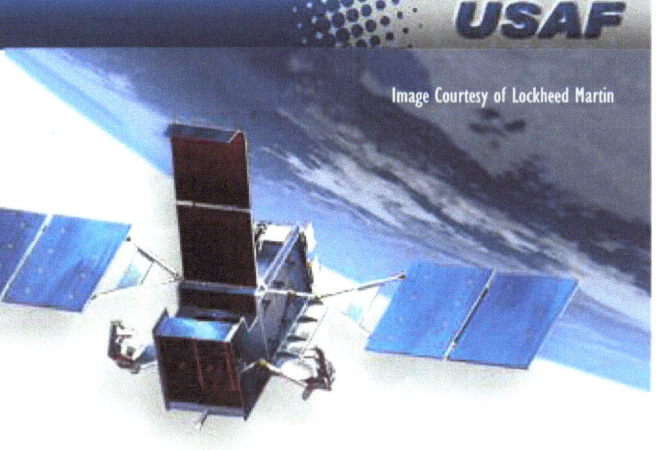

Space Based Infrared System (SBIRS) will field a constellation of satellites in Geosynchronous Earth Orbit (GEO) and hosted payloads in Highly Elliptical Orbit (HEO) with an integrated centralized ground station serving all SBIRS space elements. The SBIRS is the follow-on system to the Defense Support Program (DSP).

The infrared (IR) payload consists of:

- Scanning IR sensor two times the revisit rate and three times the sensitivity of DSP
- Staring IR sensor provides a higher fidelity and persistent coverage for areas of interest

The first HEO payload was operational December 2008. The initial launch capability for GEO-1 is on track for launch third quarter of FY 2011.

Mission: The SBIRS provides initial warning of ballistic missile launches.

FY 2012 Program: Funds the fabrication of GEO-3/4 and HEO-3/4 satellites, final integration, test, and launch of GEO-2, advance procurement of GEO-5/6, and continues ground segment fixed and mobile development.

Prime Contractor: Lockheed Martin Corporation, Sunnyvale, CA

Space Based Infrared System

	FY 2010*		FY 2011**		FY 2012 Base Budget		OCO Budget		Total Request	
	$M	Qty	$M	Qty	$M	Qty	$M	Qty	$M	Qty
RDT&E	521.5	-	530.0	-	621.6	-	-	-	621.6	-
Procurement	465.9	1	995.5	1	373.6	-	-	-	373.6	-
Spares	-	-	-	-	-	-	-	-	-	-
Total	987.4	1	1,525.5	1	995.2	-	-	-	995.2	-

* FY 2010 & FY 2011 include Base and OCO funding
** Reflects the FY 2011 President's Budget Request

Numbers may not add due to rounding

SPACE BASED AND RELATED SYSTEMS

Wideband Global SATCOM System

The Wideband Global Satellite (WGS) system is a constellation of satellites in geosynchronous orbit providing worldwide communication coverage for tactical and fixed users. Dual-frequency WGS satellites augment, then replace the Defense Satellite Communications System (DSCS) X-band frequency service and augments the one-way Global Broadcast Service (GBS) Ka-band frequency capabilities. Additionally, WGS provides a new high capacity two-way Ka-band frequency service. The WGS constellation will consist of eight total satellites and seven U.S. funded, one Australian funded.

Eight satellites

- X-band: 8 spot-beam transmit/receive via steerable phased-array antennas
- Ka-band: 10 gimbaled dish antennas
- 35 x 125 MHz channels

The fourth WGS satellite is scheduled to launch in December 2011.

Mission: The WGS constellation will provide wideband communications and point-to-point service on Ka-band and X-band frequencies.

FY 2012 Program: Provides full funding for the eighth satellite. *The sixth satellite is funded by Australia.*

Prime Contractor: The Boeing Company, El Segundo, CA

Wideband Global SATCOM System

	FY 2010*		FY 2011**		FY 2012 Base Budget		FY 2012 OCO Budget		FY 2012 Total Request	
	$M	Qty	$M	Qty	$M	Qty	$M	Qty	$M	Qty
RDT&E	67.2	-	36.1	-	12.8	-	-	-	12.8	-
Procurement	212.4	-	575.7	1	468.7	1	-	-	468.7	1
Spares	-	-	-	-	-	-	-	-	-	-
Total	279.6	-	611.8	1	481.5	1	-	-	481.5	1

* FY 2010 & FY 2011 include Base and OCO funding
** Reflects the FY 2011 President's Budget Request

Numbers may not add due to rounding

www.ingramcontent.com/pod-product-compliance
Lightning Source LLC
Chambersburg PA
CBHW050733180526
45159CB00003B/1216